Energy Storage: Legal and Regulatory Challenges and Opportunities

Author
Louise Dalton

Managing director
Sian O'Neill

Energy Storage: Legal and Regulatory Challenges and Opportunities
is published by

Globe Law and Business Ltd
3 Mylor Close
Horsell
Woking
Surrey GU21 4DD
United Kingdom
Tel: +44 20 3745 4770
www.globelawandbusiness.com

Printed and bound by CPI Group (UK) Ltd, Croydon CR0 4YY, United Kingdom

Energy Storage: Legal and Regulatory Challenges and Opportunities

ISBN 9781787422704
EPUB ISBN 9781787422711
Adobe PDF ISBN 9781787422728
Mobi ISBN 9781787422735

DISCLAIMER
This publication is intended as a general guide only. The information and opinions which it contains
are not intended to be a comprehensive study, or to provide legal advice, and should not be treated
as a substitute for legal advice concerning particular situations. Legal advice should always be sought
before taking any action based on the information provided. The publishers bear no responsibility for
any errors or omissions contained herein.

Table of contents

I.   **Introduction** 7

      1. Issues and challenges 7

      2. Current and predicted storage deployment 8

II.   **Why is electricity storage required?** 9

      1. Benefits 9

      2. Range of services 9

III. **Storage technologies** 13

      1. Technology summaries 13

      2. Technology selection 16

IV.   **Applications** 19

      1. Standalone storage projects 19

      2. Co-location with generation 20

3. Commercial behind-the-meter applications 22

4. Domestic behind-the-meter applications 24

5. Islanded networks and micro-grids 24

6. Co-location with electric vehicle charging infrastructure 25

V.   Regulatory framework 27

1. Overview 27

2. Licensing 27

3. Final consumption levies 28

4. Network charging 29

5. Definition of storage 30

6. Disputes 32

VI. Role of stakeholders 33

1. Government 33

2. Energy regulator 35

3. System operators 36

4. Network owners 38

5. Aggregators 40

6. Suppliers 41

7. Generators 42

VII. Revenue streams 43

1. Revenue stacking 43

2. Frequency response 45

3. Provision of reserve 47

4. Black Start 50

5. Alternative system services 50

6. Energy trading 50

7. Participation in balancing market 51

8. Other revenue streams 52

VIII. Electricity export and trading agreement 53

   1. Overview 53

   2. Trading arrangements 54

   3. Co-located projects 55

IX. Import electricity supply agreement 57

   1. Agreement structure 57

   2. Pricing 58

X.   Grid connection arrangements 61

   1. Key considerations 61

   2. Existing connections 63

   3. Active network management 63

XI. Construction contract 65

   1. Contractual structure 65

   2. Key issues 66

XII. Operations agreement 69

   1. Structure and key provisions 69

   2. Additional considerations 70

XIII. Land agreement 71

   1. Land requirement 71

2. Key considerations 72

XIV. Permitting 75

    1. Consent to construct and operate 75

    2. Producer responsibility 76

    3. Wider environmental and health
       and safety requirements 76

XV. Corporate arrangements 79

    1. Corporate structures 79

    2. M&A activity 80

    3. Investors 81

XVI. Financing arrangements 83

    1. Overview 83

    2. Key challenges 84

    3. Key issues 84

    4. Key financing terms 88

XVII. Conclusion 91

Notes 92

About the author 95

About Globe Law and Business 96

I. Introduction

1. Issues and challenges

The storage of electricity has the potential to solve many of the issues that the global electricity system currently faces. Until recently, the ability to store electricity at scale was limited to pumped hydro projects. However, the advent of lower-cost battery storage technologies in particular has been the catalyst for the creation of an entirely new sub-sector in the electricity industry.

As a nascent industry, the storage sector faces a variety of legal and regulatory challenges, depending on the jurisdiction, technology and application. This report provides an overview of the key issues applicable internationally in relation to the development of electricity storage projects, including:

- the principal storage technologies and their applications;
- regulatory arrangements;
- revenue streams; and
- contracting arrangements.

It also covers the key policy, commercial and legal principles that underpin this developing sector.

Energy storage has been *the* hot topic in the electricity industry in recent years. By its very nature, this report cannot be fully comprehensive or cover all issues faced by every type of electricity storage project. This report also does not cover other forms of storage, such as thermal energy storage.

2. Current and predicted storage deployment

In 2017 there were around 176.5 gigawatts (GW) of operational energy storage globally, with pumped hydro accounting for 172 GW.[1] Nonetheless, the storage sector is forecast to explode in the coming decades, with Bloomberg New Energy Finance predicting that "the global energy storage market (excluding pumped hydro) will grow to a cumulative 942GW/2,857GWh by 2040, attracting $620 billion in investment over the next 22 years".[2] This growth is expected across the globe (see Figure 1).

Figure 1. Global cumulative storage deployments[3]

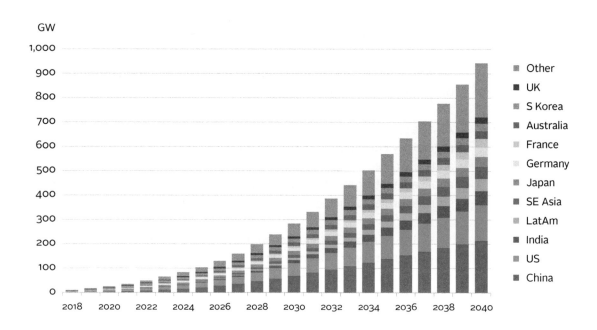

Source: BloombergNEF

II. Why is electricity storage required?

1. Benefits

Electricity storage has become a key area of focus in many jurisdictions across the globe, due to its potential to offer a wide range of benefits to electricity systems. In particular:

- the rise in intermittent, non-dispatchable generation, particularly renewables, makes it more difficult to balance electricity supply and demand and maintain the required system power quality. The rise in the deployment of intermittent technologies has further incentivised the use of storage technologies as the increasing penetration of renewables in the wholesale markets has resulted in the market becoming more liquid and more real-time activity being required;
- increased electrification of transport and heat is heightening system demand, which can increase the complexity of managing the wider electricity network; and
- in a number of jurisdictions, there is a growing lack of electricity generation capacity margin, due to the closure of old transmission-connected capacity (eg, coal and nuclear).

2. Range of services

These factors have made greater flexibility invaluable, which electricity storage technologies are ideally placed to provide. As these factors are anticipated to increase, the demand for services that can be provided by storage is likewise expected to grow.

The flexibility of electricity storage means that projects can provide some or all of the following services to the wider electricity system:

- **System services:** Storage can provide and stack ancillary services (ranging from frequency response to voltage control), creating a variety of revenue options for storage providers.
- **Balancing services:** Storage projects can participate in electricity trading, including real-time and near-real-time system balancing services, which can be particularly lucrative at times of significant system imbalance.
- **Intermittent generation integration:** Storage co-located with renewable generation enables efficient use of such generation by storing power at times of excess generation (eg, during day peaks in solar generation, when demand is insufficient) and exporting to the grid at times of peak demand and/or lower generation output. This can allow generators to optimise the price obtained and maximise the amount of electricity exported.
- **Grid reinforcement and deferral:** By reducing peak electricity demand or exported generation and providing reserve capacity, storage provides an alternative to the traditional reinforcement of existing network capacity – in particular where such reinforcement is required only for limited periods (which may be very brief or infrequent). However, the monetisation of this service has proved a significant challenge.
- **Demand reduction and peak shaving:** Storage devices can be called upon at peak times to reduce the imports of energy consumers, helping them to manage their electricity costs and reduce their demand charges, for example. Alternatively, storage can assist in increasing consumption of any relevant on-site generation by providing the flexibility to match the generation and demand profiles. The benefits of doing so will depend on the self-consumption tax and network charging arrangements in the relevant jurisdictions.
- **Price arbitrage:** Storage allows electricity to be imported and/or stored when it is cheaper and/or abundant and exported to the grid as it is needed or at times when power prices are higher. The importance and profitability of such a function will depend on the extent of price fluctuations in the relevant electricity market (ie, it plays a more important role when there is a dynamic power market with costs reflected in pricing in near to real time).

Figure 2. The range of services that can be provided by electricity storage

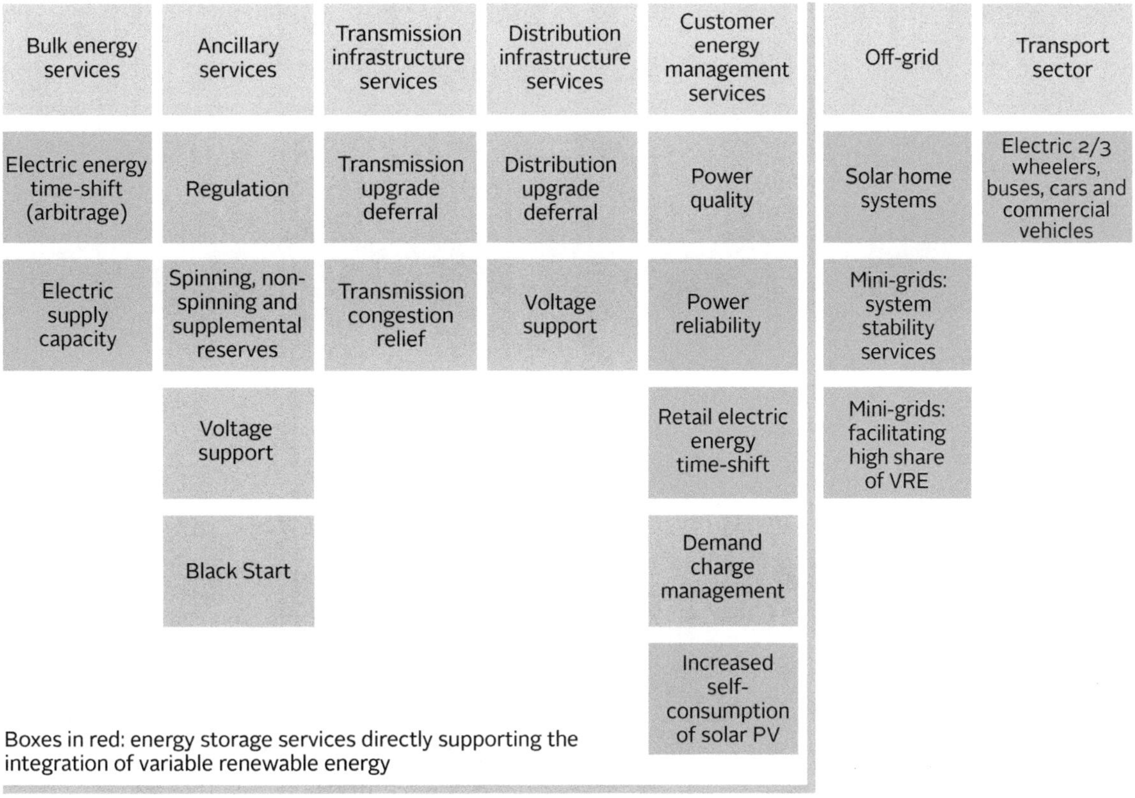

Bulk energy services	Ancillary services	Transmission infrastructure services	Distribution infrastructure services	Customer energy management services		Off-grid	Transport sector
Electric energy time-shift (arbitrage)	Regulation	Transmission upgrade deferral	Distribution upgrade deferral	Power quality		Solar home systems	Electric 2/3 wheelers, buses, cars and commercial vehicles
Electric supply capacity	Spinning, non-spinning and supplemental reserves	Transmission congestion relief	Voltage support	Power reliability		Mini-grids: system stability services	
	Voltage support			Retail electric energy time-shift		Mini-grids: facilitating high share of VRE	
	Black Start			Demand charge management			
				Increased self-consumption of solar PV			

Boxes in red: energy storage services directly supporting the integration of variable renewable energy

Source: The International Renewable Energy Agency, www.irena.org[4]

Chapter VII of this report considers the revenue stream issues in more detail.

"The International Renewable Energy Agency predicts that battery storage technology costs will fall by between 50% and 66% by 2030, due to significant anticipated deployment and further commercialisation."

III. Storage technologies

A range of energy storage technologies have been installed or are due to be deployed across the world, all with differing applications and characteristics. The two foremost technologies are currently:

- pumped hydro, which is the dominant and most mature energy storage technology; and
- lithium ion batteries, which are receiving greatest industry attention at present, in particular due to their quick reaction times, flexibility in application and significant predicted cost reductions.

1. Technology summaries

The key electricity storage technologies are summarised below. This section is not intended to be a comprehensive study of all possible technologies, as this is a rapidly moving area and the subject of much research.

1.1 Pumped hydro

The most mature form of energy storage, pumped hydro is estimated to account for 96% of currently installed electricity storage capacity.[5] It uses electricity to pump water from a lower to an upper reservoir in periods of low electricity prices or demand, which then generates power as it flows back down through turbines when required. The typical discharge times range from several hours to a few days, with an efficiency of 70% to 85%.[6] Advantages include the large capacity of such projects and their long asset lifetime. However, pumped hydro is very dependent on suitable topographical conditions, which can limit its deployment in many jurisdictions. Further, given the significant engineering requirements of such projects, there are significant upfront capital expenditure (capex) requirements and the construction timetable can be lengthy.

1.2 Electrochemical/batteries

Batteries are a key storage technology and come in a wide variety of combinations – most commonly at present lithium ion batteries. Batteries are a significant area of focus of the storage sector, due to their flexibility of application and rapid response times. In addition, significant cost reductions are anticipated, with the International Renewable Energy Agency predicting that battery storage technology costs will fall by between 50% and 66% by 2030, due to significant anticipated deployment and further commercialisation.[7]

(a) Lithium ion batteries

Lithium ion batteries are an important storage technology in both large-scale and mobile applications, such as laptops and electric vehicles (EVs). The technology is highly efficient and very flexible, due to its rapid response times and scalability. However, lithium ion batteries are susceptible to degradation, do not currently perform well at high temperatures and are not suited to longer-term storage applications. Nevertheless, there are variations depending on the relevant manufacturer, as highlighted in DNV's recent testing.[8] In addition, scrutiny is increasing around the ethical sourcing of the key raw materials for the technology, lithium and cobalt.[9]

(b) Lead acid batteries

Lead acid batteries are a well-established technology, having been invented in the 19th century. They have been used in a wide variety of applications and are relatively low cost. However, lead acid batteries have a relatively low energy density and are not well suited to full discharge.

(c) Redox flow batteries

Redox flow batteries utilise two liquid electrolyte solutions containing dissolved metal ions, which are pumped to the opposite sides of the electrochemical cell. Redox flow batteries tend to be expensive and require more space when compared with other battery technologies; however, they are less susceptible to degradation and are suitable to longer-term storage applications.

(d) Other battery technologies
Other battery technologies include:

- nickel-based batteries;
- sodium-based batteries;
- other forms of flow batteries; and
- solid-state batteries.

As illustrated in Figure 3, lithium ion batteries currently dominate new project deployments in the sector.

1.3 Flywheels
Flywheel storage functions through spinning a rotor at high speeds using electricity, creating kinetic energy that can be converted back into electricity. The main advantages of flywheels are their long lifecycle and low maintenance requirements. However, they can have variable efficiency levels and remain relatively expensive.[11]

1.4 Compressed air energy storage
Compressed air energy storage uses electricity to compress air in underground or above-ground facilities and releases it through turbines to generate power. Compressed air systems are relatively

Figure 3. Battery storage technology

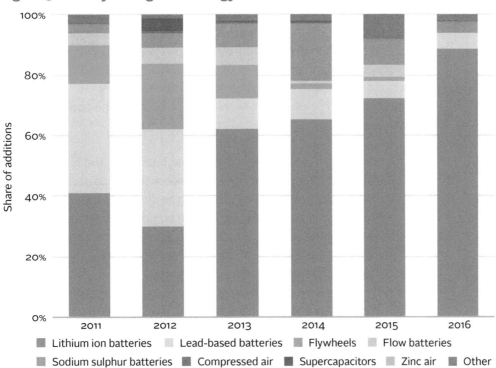

inexpensive and easily scalable (the first two projects being 321 megawatts (MW) and 110MW respectively),[12] but the air compression process can lead to significant electricity efficiency losses through the production of heat.

1.5 Hydrogen storage

Hydrogen can be an effective electricity storage process in a variety of ways – for example, utilising water electrolysis technology to produce the gas. Many industry participants are considering the production of hydrogen using excess renewable electricity.[13] The resultant gas can be stored in small amounts in tanks or in larger amounts underground, and can be used in a variety of ways, including as fuel for power production or for injection into the gas distribution network.[14] Hydrogen is scalable and can store electricity for long periods of time; however, it remains fairly expensive and not as efficient as other storage alternatives.

2. Technology selection

There is no 'one size fits all' storage technology. For example, electricity storage technologies have different power ratings (measured in MW) – that is, the power the technology can provide at any given moment – and capacity ratings (measured in megawatt-hours (MWh)), which measure the amount of electricity the technology can store. Each electricity storage technology has advantages and disadvantages, depending on a range of factors, including the required application, the relative energy and capacity, project revenue stream requirements (eg, speed of discharge) and capex and operational expenditure costs. Further, depending on the technology, significant performance improvements and cost reductions are anticipated as such technologies are fully commercialised. The choice of technology influences the applicable legal and regulatory issues, such as the following:

- **Construction requirements, including build cost and length of construction timetable:** The mass production of batteries has increased the prospects of reducing the costs of production and therefore construction of such projects. However, factors such as raw material constraints (eg, minerals such as lithium, cobalt, nickel and manganese) can serve to hinder cost reductions. Other storage technologies such as pumped hydro storage systems have exponentially higher costs when compared to batteries. A pumped hydro project requires the hollowing out of a mountain between two reservoirs in order to pump water between the two through a tunnel. Similarly, for compressed air energy storage systems, the creation of the cavern and installation of the turbines and compressor typically amount to more than 80%[15] of the initial total costs. Because of this and the

"There is no 'one size fits all' storage technology."

related ancillary requirements – such as planning consents and land rights – that must be dealt with before such developments proceed, the timetable for such projects is comparatively longer than that for the installation of other storage technologies.

- **Maintenance costs:** The costs of maintaining the different elements of pumped hydro storage systems – such as the generator, motor unit, transformers and turbines – are significantly higher compared to those for other storage technologies. The operation and maintenance (O&M) costs for batteries vary widely, as these can be highly dependent on how the battery has been operated (eg, in terms of depth of discharge and number of cycles) and therefore the extent of degradation that has occurred.

- **Speed and length of discharge:** Response times and speed of discharge for energy storage technologies vary. For example, lithium ion batteries can react in under one second.[16] The length of discharge varies between some minutes and several hours for battery technologies; whereas for pumped hydro the length of discharge varies between two hours and a few days.[17] Nevertheless, longer-term, inter-seasonal storage is yet to be delivered by any commercially viable technology.

- **Round-trip efficiency:** This is the ratio between the amount of power put into a storage device and the amount that is re-exported (measured as a percentage). No electricity storage technology is 100% efficient (ie, discharges all of the electricity which it has imported); however, round-trip efficiency can have a significant impact on the economics of a storage project, given that all lost electricity represents an operating cost. For example, compressed air energy storage systems can have a round-trip efficiency that varies between 65% and 75%, whereas lithium ion batteries can well exceed 90%.[18]

- **Asset life:** Mechanical storage technologies tend to have a longer asset life, with limited capacity degradation, when compared with most battery technologies. Pumped hydro storage systems in particular have long expected asset lives of approximately 40 to 60 years, with significant refurbishments resulting in longer lifetimes of up to 100 years.[19] Certain battery technologies can degrade substantially when utilised and therefore the number of discharge cycles and the state of charge must be carefully managed, although technology improvements are expected to seek to address this.

- **Decommissioning costs:** Due to the relative immaturity of many storage technologies, it is difficult to accurately estimate the relevant decommissioning requirements. Due to the toxic chemical residues in batteries, for example, there are currently stringent requirements in respect of battery waste in various jurisdictions. See Chapter XIV for further details.

IV. Applications

Storage is very versatile and there is a large range of potential applications. Some of the typical applications are briefly summarised below.

1. Standalone storage projects

The vast majority of storage projects that have been commissioned globally to date have been grid-scale standalone projects. This has been driven by a number of factors, including technology choice. For example, pumped hydro is not suitable for any other application.

Such projects are usually deployed in order to provide services to the wider electricity network – for example, frequency response, additional capacity in times of system stress and other system services. More recently, such projects have also begun to participate in near-real-time balancing markets in jurisdictions such as Great Britain. Depending on the jurisdiction, such projects can also benefit from price arbitrage by importing electricity when wholesale prices are low and exporting when wholesale prices are high.

Such projects have proved popular for equity and debt investors, given their relative simplicity and similarities to smaller-scale solar and other renewables projects in terms of structuring – for example, separate grid connection, land rights, construction and operation agreements and consents contained in separate special purpose vehicles (SPVs), which allow for security to be taken easily.

However, the popularity of standalone storage projects contributes to the potential downsides of this business model. The increasing number of projects seeking to participate in a specific and limited number of revenue streams has increased competition and contributed to a reduction in these revenue streams. From an overall system perspective, this competition can be viewed positively, as the costs for the consumer are reduced as a result. However, from a developer and investor perspective, the uncertainty of the future value of such revenue streams presents a challenge.

Once built and the capex paid for, the cost of running standalone storage is relatively low, comprising the cost of imported electricity, including any efficiency losses; the expenses related to addressing any degradation resulting from operating the asset; and any additional operational expenditure. This is particularly true when compared with conventional generation, which must contend with fuel, carbon and environmental costs. This allows such existing storage projects to bid into system service auctions at very low prices, further cannibalising such revenue streams for future projects.

Standalone projects' relative lack of revenue diversification, as compared to other project structures, is a risk. Further, these projects are reliant on the wider network and grid connection capacity access to provide their import electricity and ensure that they can provide the services to access the required revenue streams. This means that such projects are required to pay various connection costs and network charges. This also creates a cost exposure by requiring the project to pay at least the wholesale electricity price for the electricity it imports. In some countries these costs are even higher, as storage is treated as an end consumer. For further information on this issue, please see Chapter V.

2. Co-location with generation

The term 'co-location' covers a wide range of generation and storage project configurations. Co-located projects include:

- truly integrated solutions which are conceived, constructed and commissioned together – for example, subsidy-free solar plus storage projects;
- the retrospective addition of a storage device to an operational project; and
- standalone generation and storage projects, utilising shared land and/or grid infrastructure.

Within these types, there are various options in terms of technical configuration, such as metering, and whether the storage device is considered to be part of the generating station, is separately metered or is 'network side' within the connection point. The configuration of the generating station and the storage device directly influences the relevant issues to consider. Further, a range of corporate structures is available to such projects – for example, the same entity or different group companies owning and operating both elements or a third party owning the storage element.

There is huge potential for co-locating storage with generation, with the benefits including:

- maximising generation output and existing revenue streams, and managing intermittency and balancing costs, particularly for renewables and non-dispatchable generation;
- enabling projects to avoid grid constraint issues;
- accessing additional revenue streams – for example, the provision of frequency response – for the generation project;
- availing of possible cheaper grid connection arrangements and other cost savings associated with sharing infrastructure; and
- accessing price arbitrage and limiting exposure to negative power prices – for example, as well as the opportunity to seek to benefit from them, by getting paid to import during such periods.

This optimisation can be maximised when co-locating storage across an entire portfolio of projects or on islanded networks. Much of the interest has focused on renewables co-location, particularly to address the intermittency and balancing issues that can arise in relation to such projects. Interest in co-location is also becoming more attractive for investors in subsidy-free renewables projects – for example, Anesco's Clayhill solar plus storage project commissioned in September 2017 – due to the additional revenue streams.[20] Nevertheless, there are numerous examples of co-location of storage with conventional generation projects, from Chile and California to Australia and the United Kingdom.

A co-located project involves additional complexity and further interface risk when compared to a standalone project and, as a result, the developer or investor needs to balance this with the risk diversification and other potential upsides that co-locating storage can deliver. Such considerations also influence the decision on the relative sizing of the storage asset when compared to the generation project.

One barrier to the deployment of co-located projects is that existing renewables projects are highly likely to benefit from government

"The installation of behind-the-meter storage at commercial and industrial sites is seen as one of the largest potential growth areas for the storage market."

revenue support schemes and in some jurisdictions the impact of the co-location of storage on such renewable revenue support schemes has been unclear. However, in certain jurisdictions, regulators have sought to provide clarity. For example, in Great Britain, Ofgem has published its "Guidance for generators: Co-location of electricity storage facilities with renewable generation supported under the Renewables Obligation or Feed-in Tariff schemes", which confirms that, as long as the requirements of the Renewable Obligation Certificate and Feed-in Tariff schemes continue to be met, storage can be deployed without affecting the relevant accreditation.[21]

3. Commercial behind-the-meter applications
The installation of behind-the-meter storage at commercial and industrial sites is seen as one of the largest potential growth areas for the storage market.

The drivers from a corporate's perspective are principally cost related and can include:

- increased consumption of on-site generation, particularly in jurisdictions where renewables subsidy reductions are being implemented or proposed. This can also boost the corporate's carbon reduction and wider sustainability credentials. Self-generated electricity also has the benefit of not resulting in the corporate being charged other electricity costs, such as final consumption levies, which are charged by electricity suppliers on electricity imported from the wider network;

- managed electricity costs. This can be simply by way of increasing self-consumption as stated above, but can also involve:
  - the avoidance of peak network use of system charges, by utilising storage to reduce the site's load during the system peak. In some jurisdictions, such charges can make up a substantial proportion of the relevant corporate's electricity bill; and/or
  - access to price arbitrage capability, by avoiding high electricity import costs in more volatile markets where electricity prices fluctuate significantly and taking advantage of low electricity prices whenever these occur;
- increased energy security, by increasing the site's ability to continue to operate without being dependent on importing electricity from the wider network – for example, in system outages; and
- access to additional revenue streams through the operation of the storage device, such as grid and balancing services or demand-side response revenues.

These benefits are complemented by the range of business models available to corporates wishing to install storage. Many focus on a 'storage as a service' model, whereby the corporate does not own or operate the storage device and is simply charged a percentage of cost savings or a pre-determined fee by the storage developer. This lack of upfront investment can be attractive to corporates and, from a developer's perspective, the opportunity to contract with a creditworthy counterparty on a long-term basis enables financing options, which are further discussed in Chapter XVI.

Nevertheless, there are a variety of challenges in this sub-sector, which include the following:

- Regulatory changes can affect the potential savings available and therefore the incentive to install storage – for example, some jurisdictions are moving towards network charging models to ensure that all energy consumers contribute towards the cost of electricity networks, irrespective of the capacity or flexibility installed behind the meter;
- There is no 'one size fits all' business model and specific commercial and legal requirements need to be addressed with each corporate (eg, specific access or health and safety considerations), which adds complexity;

- There is a lack of awareness of the benefits of storage among corporates, particularly as electricity issues are not their top business priority; and
- There are practical considerations of combining the right storage technologies to fit the corporate's needs in the context of the commercial aims of the corporate and the space available at the site.

4. Domestic behind-the-meter applications

The benefits of installing storage in domestic premises broadly follows the commercial upsides outlined above, in particular in terms of maximising self-consumption of electricity generated by rooftop solar photovoltaic (PV) and reducing reliance on imported electricity. This can also unlock the potential of peer-to-peer trading utilising blockchain technology, which is currently being explored in a number of jurisdictions, including the United States and the United Kingdom.[22]

In jurisdictions where there is significant domestic solar installed, such as Germany, the domestic storage market is growing rapidly, with over 100,000 households with batteries installed as of August 2018.[23] In addition, some governments are specifically choosing to incentivise such deployment by providing grants or other subsidies (eg, in Ireland).[24]

However, the upsides for domestic users are not as varied or as extensive as with commercial applications, which means that there is a longer payback time for the initial capex for the storage device. The impact of self-consumption taxes and the potential negative impact on applicable renewable export tariffs can further undermine the case for installing storage for domestic consumers.

5. Islanded networks and micro-grids

Remote regions and islands face specific challenges when it comes to maintaining adequate generation and balancing supply and demand. The lack of interconnection with wider networks means that such systems must be self-sufficient and independent. This has often limited the ability to deploy a significant amount of intermittent generation in such regions, given the inherent lack of reliability and dispatchability of such technologies. Further, due to the relatively low demand and fuel supply issues in such areas, there are limited opportunities to deploy highly efficient and cleaner conventional generation, such as combined cycle gas turbine projects. As a result, the deployment of storage can have a multitude of benefits:

- It allows for greater penetration of intermittent renewables – for example, significant deployment of solar on tropical islands, enabling cleaner electricity generation; and

- It can reduce expensive and carbon-intensive fuel costs, such as diesel.

Some governments have run specific solar plus storage tenders for remote islands – for example, in 2017 France ran a successful tender resulting in almost 64MW of projects in Corsica, Guadeloupe, Guyana, La Réunion, Martinique and Mayotte.[25]

The technology choice is important in islanded/micro-grid applications, as if the intention is to power an island network solely with solar, for example, the storage technology must be suited to longer-term, overnight discharge.

6. Co-location with electric vehicle charging infrastructure
The installation of storage alongside EV charging points has a number of potential advantages, particularly in relation to the grid connection element, which can represent the most significant capex cost for individual charging points at present. The co-location of storage can lessen the impact of EV charging on the wider network by reducing the peak demand of the charging point, which can be significant, particularly in the case of super-rapid charging points. This can decrease the risk that wider network reinforcement is required,

"The installation of storage alongside EV charging points has a number of potential advantages, particularly in relation to the grid connection element, which can represent the most significant capex cost for individual charging points at present."

which can reduce the cost of the charging point grid connection itself or limit any delay to the delivery of the grid connection.

Further, by levelling out the demand of EV charging points, storage can provide benefits in terms of system stability and provide additional revenue streams. To date, deployment has been mainly focused on battery technologies, given their rapid response time and relative space efficiency. Nevertheless, EV charging infrastructure itself faces various issues, such as siting and demand risk, and the cost of charging remains a merchant revenue. As a result, the additional cost of co-locating storage needs to be carefully considered in light of the proposed benefits. This report does not explore the various legal and regulatory issues in relation to the significant vehicle-to-grid opportunities presented by the mass roll-out of EVs, in part due to these opportunities being in the test and demonstration phase.

V. Regulatory framework

1. Overview

Most developed electricity markets have well-established electricity regulatory regimes, which distinguish between electricity generation, supply and electricity networks in both primary and secondary legislation and the relevant electricity industry codes. Invariably, however, the regulatory position of storage has been unclear. This is a result of a general lack of definition of the term 'storage' in key national regulations and therefore uncertainty as to whether it should be treated as an existing category of stakeholder – usually generation, demand or both – or should be assigned its own categorisation, recognising the unique features of storage technologies. As further discussed below, the lack of clarity in relation to the definition and treatment of storage can have various consequences, such as the role that various industry participants can play in relation to storage. While the lack of a definition has not in itself prevented the deployment of energy storage, there are various unintended consequences resulting from this lack of clarity, which include the following.

2. Licensing

The lack of clarity on how storage should be classified can result in confusion about which types of energy storage projects do or do not require a licence in the relevant jurisdiction and which type of licence is required if so. In most jurisdictions where licences have been

"The lack of clarity on how storage should be classified can result in confusion about which types of energy storage projects do or do not require a licence in the relevant jurisdiction and which type of licence is required if so."

required, these have been a form of generation licence, often due to the historic treatment of existing pumped hydro assets. Alternatively, storage currently falls completely outside the current licensing regime.

This can have far-reaching implications – for example, operating without a required licence or benefiting from a relevant exemption from a requirement to hold a licence can be an offence. Further, a project's licence status can define its required compliance with and treatment pursuant to the applicable industry codes and various charges, as further illustrated below.

3. Final consumption levies

Where a storage project imports electricity from the wider system via an electricity supply agreement, it may be charged final consumption levies – which are designed to be charged on final demand consumers, to fund renewables subsidies and environmental programmes, for example – on its imported electricity. However, most of that electricity imported into a storage project is re-exported to the system (depending on the efficiency of the relevant technologies). This creates a distortion in the market for storage operators by increasing operational expenditure and recouping charges that were never intended to be recovered from such projects.

There is a range of options to resolve such unequal treatment. More complex options provide relief from such levies in respect of electricity that is subsequently re-exported back to the system, which brings the efficiency of the relevant storage technology into focus, but requires potentially complex metering and reporting. A simpler approach is to exempt storage projects from paying such levies in their entirety, which can be achieved by way of a definition of 'storage', licensing or some other means.

4. Network charging

The confusion about whether to treat storage as generation or demand or its own separate category flows through into the costs paid for utilising and balancing the wider electricity network. Often, this has led to storage being treated as either generation or demand – which may even differ depending on whether the project is connected at a high or low voltage – or being double charged as demand when it imports electricity and as generation when it exports. Such charging has led to confusion among developers and created potential higher barriers to entry as a result. Again, this inequality has been identified in many jurisdictions and there are various proposals to resolve it.

Case study: Great Britain
Licensing: In 2016 the government and the regulator, Ofgem, published a call for evidence which considered, among other things, how storage should be treated for the purposes of licensing.[26] The options included:

- continuing to treat storage as generation for licensing purposes (ie, the existing approach);
- defining storage as a subset of generation in a modified generation licence specifically for storage;
- defining storage in primary legislation as a subset of generation in the Electricity Act 1989 with a modified generation licence for storage; and
- defining storage as a new activity with a separate storage licence.

Following this consultation, on 29 September 2017 Ofgem published a licence consultation on modifying the electricity generation licence to accommodate electricity storage.[27] This licence consultation was accompanied by a proposed update to the standard generation licence conditions so that they would apply to energy storage providers as well as electricity generation providers.[28] However, while it was expected that these amendments would be in place by Summer 2018, as of May 2019 the statutory consultation to implement the changes had not been launched.[29]

In the meantime, projects can continue to hold a traditional generation licence and the existing class licence exemptions remain available.

Final consumption levies: The government and Ofgem have stated that storage facilities in Great Britain holding a licence should not be liable for final consumption levies. This means that smaller-scale projects must hold a licence, with its attendant obligations (as opposed to a small-scale exemption). Further, in order to ensure that such projects are exempt from certain levies, further industry code modifications are required, which are currently in train.[30]

Network charging: In 2017, Ofgem stated that:

- storage should be treated as generation for the purpose of setting all residual charges, and should not pay demand residual charges for either transmission or distribution;
- it is appropriate for storage projects to pay 'forward-looking' charges that reflect the incremental demand and generation costs that such projects place on networks; and
- storage should not pay a disproportionate amount for balancing use of system charges compared to other forms of generation.

Nevertheless, the regulator has looked to industry to propose code modifications to resolve these issues. At the time of writing, these industry modifications have not yet been implemented.[31]

5. Definition of storage

As has been illustrated above, the approach to the widespread adoption of storage has not been consistent in the key jurisdictions where new storage projects are being deployed. This is perhaps best demonstrated by the range of definitions currently proposed in such jurisdictions of exactly what it comprises. The challenge with defining storage is the extent to which a definition can be as technology neutral as possible, while still being sufficiently precise to provide the clarity required by industry participants. Common challenges include the following:

- Does the definition treat storage as a subset of generation or demand or as an asset class in and of itself, recognising the distinct characteristics of storage?
- Does the definition encompass capacitors and transformers, which are traditionally network assets? If so, this could lead to issues in jurisdictions where legal unbundling of networks has been implemented.
- Does the definition seek to cover electricity storage only or thermal technologies as well?
- What size and scale of storage projects should be captured

by the definition? For example, should there be a *de minimis* threshold? Will this apply to EVs providing vehicle-to-grid services?

In fact, a very limited number of jurisdictions have thus far defined energy storage within their legislative frameworks.[32]

In the United States, Federal Energy Regulatory Commission (FERC) Order 841 defines 'electric storage resource' as: "a resource capable of receiving electric energy from the grid and storing it for later injection of electric energy back to the grid." [33] This definition applies to all electricity storage technologies, but not thermal storage, and to electricity storage projects connected at a transmission and distribution level. It also applies to behind-the-meter projects where the project exports electricity to the wider network.

In Europe, under the Clean Energy for All Europeans Package (also known as the Winter Package), the proposed revised Electricity Directive includes the following definition: "'energy storage' means, in the electricity system, deferring an amount of the electricity that was generated to the moment of use, either as final energy or converted into another energy carrier." [34] This definition is broad,

"The challenge with defining storage is the extent to which a definition can be as technology neutral as possible, while still being sufficiently precise to provide the clarity required by industry participants."

encompassing power-to-x and thermal systems. Nevertheless, the Electricity Directive will need to be transposed into national law, so the impact or interaction with existing legislation in the various EU jurisdictions is not currently clear.

In the Irish single electricity market, no overall definition of 'storage' has been introduced into the industry codes; however, specific definitions of 'battery storage unit' and 'pumped hydro unit' have been included.[35] Both technologies are considered to be 'generator units'.

In Great Britain the government has stated that it intends to introduce the following definition into primary legislation when parliamentary time allows: "'Electricity Storage' in the electricity system is the conversion of electrical energy into a form of energy which can be stored, the storing of that energy, and the subsequent reconversion of that energy back into electrical energy."[36] In the meantime, storage is undefined in primary legislation in Great Britain and is being classed as a subset of generation for licensing purposes (for further details see the case study on p29.)

6. Disputes
One area that has yet to be fully explored is the treatment of different storage technologies under international dispute resolution treaties, such as the Energy Charter Treaty.[37] As international investments and therefore disputes increase in the future, it is expected that further clarity will be forthcoming.

VI. Role of stakeholders

1. Government

National, federal and regional governments play an important role in providing positive signals to the storage industry, which can include:

- the direct procurement of energy storage projects;
- the introduction of targets for energy storage to be met by relevant industry participants;
- the funding of research and development into the development of different types of energy storage technologies – for example, Highview Power's ground-breaking grid-scale liquid air energy storage demonstration project has benefited from government funding and grants;[38] and
- consultation on and implementation of measures to address issues and barriers to entry in the regulatory framework, such as those explored in Chapter V.

Case study: South Korea – co-location promotion
The South Korean government has promoted the installation of energy storage alongside renewables by assigning a higher weighted ratio to the number of renewable energy certificates issued to energy storage devices associated with wind and solar projects than to projects without storage installed.[39] This resulted

in nearly 1.1 GWh of energy storage being deployed in South Korea in 2018.[40]

Case study: Jordan – direct government procurement

In July 2017 the Ministry of Energy and Mineral Resources of Jordan launched a tender for the procurement of energy storage systems in the Kingdom, the first phase of which is intended to comprise a 30MW/60MWh electricity storage plant connecting a substation in Ma'an, which is in the vicinity of a number of solar PV plants. The initial project, which has an anticipated project life of 15 years, is intended to provide the following services:

Ramp-rate control of PV and Wind power plants in the Ma'an area to smooth the power output.
Reduction of conventional power plant operation necessary for spinning reserve.
Energy shift of otherwise curtailed renewable energy to times of peak demands.[41]

A large number of companies have been shortlisted for the tender, but a preferred bidder has yet to be announced.[42]

Case study: California, United States – industry-mandated procurement

In October 2010 California enacted a law requiring the California Public Utility Commission (CPUC) to establish appropriate 2015 and 2020 energy storage procurement targets for electricity corporations, if cost effective and commercially viable, by October 2013.[43] In October 2013 the CPUC set a mandate for utilities PG&E, Edison and SDG&E to procure 1,325MW of energy storage across the transmission, distribution and customer levels by 2020, with the projects required to be delivered by no later than the end of 2024.[44] As a result, as of October 2018 California's three largest investor-owned utilities had procured 1,620MW of energy storage.

However, it is noteworthy that only around one-third of the capacity procured has been commissioned.

Further, in September 2018 California passed Senate Bill 100 (SB 100), which puts California on a pathway to achieve 100% zero carbon electricity by 2045. This goal is to be reached in a variety of ways, key among which is through enhancing its storage capabilities so as to make better use of intermittent power-generating technologies.[45] In a further demonstration of support for the energy storage sector, the California legislature has enacted two additional bills that complement SB 100, as follows:

- Senate Bill 700, adding more than $800 million in funding for behind-the-meter storage and extending the Self-Generation Incentive Programme (SGIP) until 2026. To date, the SGIP has contributed to the delivery of over 300MW of behind-the-meter energy storage;[46] and
- Senate Bill 1369, for the first time defining green electrolytic hydrogen as an eligible form of energy storage.[47]

2. Energy regulator

As has been made clear, global electricity systems have not been designed with energy storage in mind. The role of the regulator is to assist in removing barriers to entry to energy storage projects and ensure a level playing field. This may include:

- reforming licensing regimes, as further described in Chapter V, section 2;
- providing clarity in relation to eligibility of co-located renewables projects for renewable subsidies;
- defining the role of existing stakeholders in relation to storage, in particular networks. For further discussion, see Chapter VI, sections 3 and 4; and
- directing the industry-led amendment of, or directly amending, industry rules and network codes.

"Global electricity systems have not been designed with energy storage in mind. The role of the regulator is to assist in removing barriers to entry to energy storage projects and ensure a level playing field."

Case study: United States

In February 2018 FERC issued Order 841, which aims "to remove barriers to the participation of electric storage resources in the capacity, energy and ancillary services markets operated by Regional Transmission Organizations ('RTOs') and Independent System Operators ('ISOs')".[48] Order 841 aims to increase competition, ensure that RTO and ISO markets produce "just and reasonable rates", and increase wider electricity system resilience by providing clarity around market access and clarifying existing industry rules.

The specific requirement of Order 841 is that each RTO and ISO provide a tariff that establishes a "participation model" for energy storage, which:

- ensures that energy storage is eligible to provide all capacity, energy and ancillary services that it is technically capable of providing;
- allows energy storage to set the wholesale market clearing price on both import and export consistent with existing market rules;
- recognises the physical and operational characteristics of energy storage through bidding parameters or other means; and
- establishes a minimum size requirement that does not exceed 100 kilowatts (kW).

In addition, the import of electricity from an RTO or ISO market for energy storage, which the project then re-exports back to that market, must be charged at the wholesale locational marginal price.

RTOs and ISOs are required to submit their compliance filings establishing their participation model in December 2018, with the deadline for implementation the following year.

However, Order 841 does not address all elements of the energy storage market – in particular, in relation to network charging and distributed energy storage issues. In addition, some commentators have noted that Order 841 gives ISOs a significant amount of flexibility in terms of implementation.

3. System operators

System operators are responsible for planning and managing the day-to-day operation of the relevant electricity system. In particular, a system operator is responsible for:

- system stability and security; and
- balancing of demand and supply of electricity in real time.

"In order to fulfil its role, a system operator will procure a range of services from generators, consumers and storage providers. Such services include existing ancillary and reserve services, and services specifically designed or well suited to flexible and storage technologies."

This includes national and regional system operators, as well as those at different voltage levels such as transmission and distribution.

In order to fulfil its role, a system operator will procure a range of services from generators, consumers and storage providers. Such services include existing ancillary and reserve services, and services specifically designed or well suited to flexible and storage technologies. The increase in demand for such services as part of the energy transition creates opportunities for electricity storage providers. As further discussed in Chapter VII, such services provide essential revenue streams for storage developers. As a result, there has been demand for:

- clarification in terms of providing a number of these services at the same time (so-called 'stacking');
- clarification on who is eligible to provide such services;
- removal of barriers to participation by storage in such markets; and
- greater transparency in terms of future procurement needs and assessment of bids for such services.

For example, the EU Clean Energy for All Package requires that the procurement of non-frequency ancillary services be transparent, nondiscriminatory and market based, ensuring effective participation of energy storage facilities.

> **Case study: Great Britain**
> In June 2017 National Grid, the Great Britain system operator, published the System Needs and Product Strategy (SNAPS) consultation, in order to improve balancing services and markets by providing greater clarity and investor certainty. SNAPS summarised future Great Britain electricity transmission system needs in five key products – inertia, frequency response, reserve, reactive power and Black Start (see Chapter VII) – and considered how these and associated information provided could be improved.[49] This has resulted in the publication of specific product roadmaps outlining the specific actions in relation to each product in order to achieve these aims.[50]

4. Network owners

Conceptually, electricity network owners are well placed to leverage the benefits that storage can provide – for example:

- utilising knowledge of the system to ensure optimal location, sizing and characteristics of the storage project to avoid network reinforcement; and
- integrating storage into long-term, system-wide resource planning to procure an efficient system and reduce network costs, which are ultimately passed onto consumers and currently make up a significant proportion of the average consumer's electricity bill.

However, there are concerns that networks owning and operating storage have the potential to create conflicts of interest between the network's core functions, potentially expanding the monopoly power of the network. Further, there are concerns that network involvement could distort or even foreclose the wider flexibility market for third-party providers.

To date, there have been differing approaches across the world on the role of network owners in relation to networks owning and operating storage. For example:

- in California, the CPUC allows for utility ownership of energy storage, but this is limited to 50% of storage located within each segment of the system: transmission, distribution and customer-side. This approach has been designed to encourage utilities to procure storage from a range of ownership structures, including third-party and joint ownership of systems;[51] and

- through its Reforming the Energy Vision initiative, the New York Public Service Commission has limited utilities from owning storage other than in the following circumstances:
  - where a procurement has been solicited to meet a system need and a utility has demonstrated that competitive alternatives proposed by non-utility parties are "clearly inadequate or more costly";
  - where energy storage and generation is sited on utility property;
  - where markets are not adequately serving the needs of low-income communities; and
  - for demonstration projects.[52]

The situation is complicated further in jurisdictions that are subject to unbundling principles, which are most recently embodied in the EU Third Package legislation. This restricts the involvement of entities with interests in the transmission and distribution of electricity from also being involved in electricity production or supply. Nevertheless, the approach of various jurisdictions has not been entirely consistent:

- In Italy, the legislative framework allows the transmission system operator, Terna, and distribution network operators (DNOs) to

"There are concerns that networks owning and operating storage have the potential to create conflicts of interest between the network's core functions, potentially expanding the monopoly power of the network. Further, there are concerns that network involvement could distort or even foreclose the wider flexibility market for third-party providers."

develop and manage storage facilities, provided that the investment in storage can be justified when compared with the alternative approach.[53]

- In Great Britain, by contrast, Ofgem has proposed an additional licence condition to ensure that, regardless of their size, DNOs cannot operate storage facilities or small-scale generation.[54] The licence condition applies to all storage and generation, whether or not such assets are required to have a generation licence. However, where the DNO meets the obligations under Great British and EU legislation to legally separate the operation of storage or generation assets, this will be permitted. In addition, there are certain exceptions to the licence condition in relation to island network generation, where a DNO operates existing storage facilities, network safety and where the DNO applies for a specific exemption from Ofgem.

This approach is similar to that being taken at an EU level under the Clean Energy for All Europeans Package. The proposed revised Electricity Directive,[55] which had reached political agreement at the time of writing but had not yet been adopted, proposes that "distribution system operators shall not be allowed to own, develop, manage or operate energy storage facilities". However, specific derogations allow member states to do so if the following conditions have been fulfilled:

- Following an open and transparent tendering procedure, third-party entities have not expressed their interest to own, develop, manage or operate storage facilities;
- Storage is necessary for the distribution system operators to fulfil their obligations under the directive for the efficient, reliable and secure operation of the distribution system; and
- The relevant regulatory authority has granted its approval.

5. Aggregators

Aggregators can play an important role in allowing energy storage projects to access the balancing and ancillary services markets. However, contracting through aggregators brings challenges, but can also yield significant advantages, including:

- access to markets that would otherwise be unavailable to smaller-scale distributed storage assets due to their size and location;
- the ability to leverage established relationships of the aggregator with the relevant contracting authority, such as the relevant system operator;
- the potential to increase revenue and reliability and reduce

"The key challenge for project finance is the credit and performance risk of interposing an aggregator into the structure, which may not have a healthy balance sheet. Careful consideration must be given to insolvency risk, the aggregator's technology and intellectual property, and its customer base (in cases where a battery is aggregated with third-party assets)."

utilisation and degradation risk through aggregating storage, particularly batteries with other assets; and

- the use of sophisticated control systems to prioritise dispatch to capture maximum revenues.

The key challenge for project finance is the credit and performance risk of interposing an aggregator into the structure, which may not have a healthy balance sheet. Careful consideration must be given to insolvency risk, the aggregator's technology and intellectual property, and its customer base (in cases where a battery is aggregated with third-party assets). Further, where an aggregator installs equipment and controls the dispatch of a battery, there is an interface risk and any agreement with an aggregator must work with the construction and O&M contracts, as well as any warranty provided by the battery manufacturer. While it is not possible to eliminate these risks entirely, a number of structures are possible.

6. Suppliers

Licensed electricity suppliers have a variety of roles in relation to storage projects – principally:

- the supply of import electricity to such projects, as discussed further in Chapter IX;

- the offtake of electricity exported by such projects and access to trading markets, as discussed further in Chapter VIII; and
- the offer of storage to domestic and non-domestic customers, as part of their demand-side response or wider energy efficiency services.

7. Generators

The main role of generators in relation to electricity storage concerns potential co-location opportunities, as discussed in Chapter IV. Nevertheless, there is a range of other roles that generators could play – for example, private wire or other direct supply arrangements for the import electricity of such projects.

VII. Revenue streams

1. Revenue stacking

Given that, in the main, there are no specific subsidies designed to enable storage in most jurisdictions, storage projects focus on the revenue streams that the relevant technology can provide. However, there is unlikely to be a single revenue stream that alone is sufficient to justify the development of a storage project. As a result, storage projects are creating investable business cases by seeking to stack revenue streams – specifically, a combination of the provision of system services, balancing and/or reserve.

In contrast to conventional generation projects, where the sale of electricity makes up a significant proportion of the revenue stream, in the case of storage, electricity sales alone may not meet the costs of the import electricity due to efficiency losses and potential final consumption levies and double charging issues, as illustrated in Chapter V.

Revenue stacking provides a project with greater revenue stream resilience, rather than being reliant on one or two revenue sources, which may be subject to reductions due to overcapacity in the market, fluctuations in payments and regulatory reform. Nevertheless, a reliance on revenue stacking presents storage projects with various challenges. First, each specific system service often tends to have:

"Revenue stacking provides a project with greater revenue stream resilience, rather than being reliant on one or two revenue sources, which may be subject to reductions due to overcapacity in the market, fluctuations in payments and regulatory reform."

- different tender frequencies, ranging from daily or weekly auctions in jurisdictions such as the Netherlands[56] to annual or ad hoc tenders. The frequency of the tender tends to correlate to the length of contract and how real time the auction is;
- different tender timelines – that is, the point in the development cycle at which a project is eligible to participate in the tender and the development criteria it needs to demonstrate at that time. For example, where the tender is designed to facilitate new build projects, this can result in projects having a gap prior to the start of such revenue stream that they must fill;
- different availability windows and contract lengths, which can range from one week to a matter of months or a number of years;
- different payment structures, with combinations of availability and utilisation payments and other optional fees;
- different technical requirements, such as minimum capacity thresholds, response times, metering requirements and technical availability; and
- different counterparties – while most commonly this may be the system operator or network owner, there can be a range of counterparties with differing terms and conditions and varying creditworthiness.

However, there has been criticism over the level of transparency in relation to the size of the flexibility opportunity, including the technical requirements of the service, how much capacity will be procured during each tender and the considerations that will be taken into account in the procurement, such as availability price, utilisation price, location, capacity and the identity of the developer.

In addition, often the terms and conditions for the relevant services have not been designed to allow for revenue stacking – for example, by intentionally or unintentionally providing for exclusivity of provision for one service in the relevant tender period. All of these factors need to be carefully managed in order for the revenue streams to be successfully delivered and the project to avoid penalties for non-delivery. A further barrier to internationalisation of such business models is that these elements are market specific.

Indeed, some system services are not currently openly tendered, but rather bilaterally negotiated or awarded by the relevant system operator or other contract counterparty. This has been a perceived barrier to entry for storage, which can often be smaller scale and developed by newer market entrants when compared with the traditional participants in such markets. Nevertheless, in the majority of jurisdictions, the number of providers of system services has grown dramatically in recent years due to the energy transition, illustrating the increasing number of new entrants.

A further issue is that the terms and conditions of such services often have not been designed to be a core income stream for such projects. This means that the bankability of the terms can leave something to be desired – for example, including hair-trigger termination events.

2. Frequency response
Frequency response is a service provided to the network to help maintain system frequency (ie, the real-time balance between generation and demand on the system) within the limits required by the relevant industry code. Storage is particularly well suited to providing frequency response, as it can act as both demand and generation, and can therefore respond to both frequency increases and decreases and – depending on the technology – respond rapidly.

Frequency response is one of the main system services that has been the focus of storage developers to date. However, as with all such system services, frequency response is a finite market in each jurisdiction and therefore the prices paid and the amount procured have both been reducing. Further, in many jurisdictions there are ongoing reforms of the procurement of frequency response, in part to reflect the energy transition, the advent of storage as a viable provider of such services and the move to a smarter system – for example, by way of closer to real-time procurement of the service. In addition, the ability to provide frequency response and trade the capacity separately can be restricted, which limits the potential revenue stack available during the availability windows that the project has contracted to provide frequency response.

Case study: Great Britain – enhanced frequency response[57]
The enhanced frequency response (EFR) tender run by system operator National Grid in Summer 2016 has been widely heralded as kick-starting the deployment of standalone battery storage projects in Great Britain.

The EFR tender was hugely popular, with 64 companies pre-qualifying over 1.2GW of projects. While the tender was technology neutral, the process attracted significant interest from energy storage developers due to the characteristics of the new service, which required the service provider to deliver output within one second (or less) of registering a frequency deviation.

National Grid imposed a 50MW cap per bidder in order to ensure that a competitive market developed for EFR. As part of the assessment criteria, bidders were required to:

- provide details of the site location;
- provide a programme of works;
- demonstrate a valid connection offer for the site (with a connection date no later than 18 months from contract award);
- demonstrate an option, a lease or a freehold interest in the site;
- demonstrate that financing was in place (either internally or by way of a letter of intent from a funding party); and
- demonstrate that they had either contracted with or received bids from an engineering, procurement and construction (EPC) contractor or equipment manufacturer.

In total, National Grid received 37 bids by close of bidding on 15 July 2016.

The EFR contract provides service providers with availability payments for provision of the EFR service for four years from satisfaction of the conditions precedent, which include passing a commissioning proving test. The EFR tender was very popular and therefore produced pricing far lower than was initially expected, at between £7/MWh and £12/MWh. This pricing put pressure on projects to secure other revenue streams, such as participation in the capacity market, and seek cost savings. It is estimated that this new service will result in reduced system operation costs of approximately £200 million.

In total, eight lithium ion battery storage projects with a total capacity of 201MW secured EFR contracts in the tender, as set out in Table 1.

Table 1. EFR contract winners

EFR winners	MW
EDF Energy Renewables	49
Vattenfall	22
Low Carbon	10
Low Carbon	40
E.ON UK	10
Element Power	25
RES	35
Belectric	10
Total MW	**201**

Initially, two further EFR tender rounds – each for an additional 200MW of capacity – were expected. However, National Grid announced in June 2017 that no further EFR tenders would be held, with the service to be incorporated into wider response products procured by the system operator.[58]

3. Provision of reserve

The provision of additional capacity to the system by storage assets can help to manage situations where system demand exceeds supply. A variety of such services can be provided, which broadly fall into two categories:

- preventive – the provision of additional reliable capacity to the system (by way of either additional output or reduction in demand) in anticipation of a reduction in the available capacity margin in order to ensure the long-term security of electricity supply. This includes capacity markets; or

- reactive – the provision of additional capacity in short order when demand on the system is greater than was originally forecast or unforeseen generation unavailability has occurred.

The nature of the service can influence the terms on which the reserve service is contracted. For example, new build capacity is often contracted on different terms to reflect the risk of non-delivery of the reserve when required and also the requirement for upfront capital investment. This often results in longer-term contracts, but with additional monitoring and milestones during the development period in order to ensure that the capacity will be forthcoming.

Nevertheless, the time-limited duration of all electricity storage technologies for the provision of reserve can have an impact on the suitability to provide reserve. In particular:

- the state of charge of the storage projects may not be optimal when the reserve requirement occurs – for example, if the project is simultaneously participating in multiple commercial services as part of its revenue stack;
- as discussed above, certain storage technologies such as batteries are subject to degradation, which could limit the

"New build capacity is often contracted on different terms to reflect the risk of non-delivery of the reserve when required and also the requirement for upfront capital investment. This often results in longer-term contracts, but with additional monitoring and milestones during the development period."

ability of the project to adequately provide the required duration of reserve; and

- the relevant reserve service may require sustained capacity, which could be longer than the total technical duration of the relevant energy storage project.

The impact of such time-limited storage on the provision of reserve can be mitigated in a variety of ways, such as the following:

- the assignment of de-rating factors to different technologies in order to recognise that a project may not always be available to provide the relevant reserve due to the factors identified above;
- periodic testing to demonstrate that the project has the technical capability to provide the reserve when required;
- performance monitoring where the service is called upon; and
- provision for penalties for non-delivery of reserve and, potentially, failure to pass the relevant tests; however, the levels of such penalties will influence the effectiveness of this deterrent.

Case study: close to near time-procurement – Project TERRE
There are also potential future reserve revenue streams, such as Project Trans European Replacement Reserves Exchange (TERRE).[59] Project TERRE is the European implementation project for exchanging replacement reserves in line with EU Regulation 2017/2195 establishing a guideline on electricity balancing, which came into force in December 2017.[60] The objective of Project TERRE is to implement a multi-system operator platform to obtain all replacement reserves offers and to optimise the allocation of such reserves internationally across the relevant jurisdictions. The participating countries are France, Great Britain, Italy, Portugal, Spain, Switzerland, Czech Republic, Poland and Romania.[61] This will represent a new international market for storage and has the potential to increase liquidity and opportunities to offer flexibility.

Generation, demand-side response and storage will be eligible to participate either directly, provided that the asset is at least 1MW, or via an aggregator. There will be hourly auctions comprising blocks of 15-minute periods, which will be 'pay as clear'.

The 'go-live' for Project TERRE is planned in December 2019. It is expected that Project TERRE will reduce balancing costs across the participant jurisdictions. However, the constraints that the current level of interconnection will place on the operation of the market are currently unclear.

Separately, the Project Manually Activated Reserves Initiative – which aims to establish a platform that balances energy from frequency restoration reserves with manual activation – is also being developed.[62]

4. Black Start

Black Start is the system service used to restore power in the event of a total or partial shutdown of the relevant electricity system. This potentially lucrative service has traditionally been provided by conventional generation. To date, there have been limited examples of the widespread use of storage to provide the service; however, there have been demonstrations that battery storage is capable of providing the service in Germany[63] and California.[64] Given the unique nature of Black Start – for example, the requirement for a geographically disparate provision and its vital system security role – an open competitive procurement process is challenging. Nevertheless, it is expected that storage projects will seek to provide such service where possible in the future.

5. Alternative system services

Storage assets are capable of providing a number of other services, such as reactive power, which is the service by which system operators ensure that voltage levels on the system remain within a specified range. Storage is well placed to provide such service, as it can both absorb reactive power (decreasing voltage) and generate reactive power (increasing voltage), which helps the network to perform more efficiently by minimising losses. Further, it is expected that future revenue streams may become available as markets move to a distribution system operator model – for example, monetisation of the avoidance of network reinforcement.

6. Energy trading

Trading on the wholesale electricity market occurs prior to the delivery of the electricity in real time. This is over various time periods, from years or seasons ahead up until shortly before physical delivery. Trades can be executed:

- through power exchanges, where offers for specified volume of generation at a specified price are matched with bids for demand; or
- bilaterally between generators and suppliers, known as over-the-counter trades.

The key electricity trading markets for storage are nearer real-time markets, often known as the day-ahead and intra-day markets, as there is an increasing value for flexibility the closer to real time the trading occurs. Nevertheless, the closer to real time that trading

"It is expected that future revenue streams may become available as markets move to a distribution system operator model – for example, monetisation of the avoidance of network reinforcement."

occurs, the more difficult it can be to forecast the prices that can be achieved. As a result, the trading strategy, including the risk appetite of the developer, will influence the proportion of capacity traded in each market. Such an approach brings further challenges in terms of requiring more active management of the storage device and presenting state of charge and degradation risks for certain technologies, such as batteries.

The mechanics of accessing such trading markets are discussed further in Chapter VIII.

7. Participation in balancing market
While trading in the wholesale market seeks to match the level of predicted demand and supply ahead of time, differences between the forecasted and actual volumes delivered are inevitable. As a result, system operators utilise balancing mechanisms in order to ensure that supply and demand are kept in equilibrium in real time.

Given the immediate requirement, prices for participation in such balancing mechanisms tend to be higher than those in earlier wholesale market trading. Storage is well placed to capitalise on such volatile pricing. Nevertheless, there can be various barriers to entry for the participation of smaller-scale, distributed storage in such mechanisms – for example, requirements for a specified minimum capacity, accession to relevant codes, the need for specific equipment and other obligations which result in high cost of participation and compliance.

The mechanics of accessing such balancing markets are discussed further in Chapter VIII.

8. Other revenue streams

The above sets out some of the key revenue streams available to storage, focusing on the grid-scale applications. However, a variety of others may be available depending on the jurisdiction, such as the following:

- Revenue streams may be available by virtue of the structure of the network charging in the relevant jurisdiction. For example, in Great Britain, until a recent code modification curbed their value, distribution-connected storage could receive payments of £45/kW for exporting during 'Triads' (the three half-hour settlement periods each winter when electricity demand in Great Britain is highest, separated by 10 clear days), as a result of reducing the net demand on the distribution network; and
- Depending on the jurisdiction, various tax incentives may be available for specific storage projects. However, consideration of tax is outside the scope of this report.

VIII. Electricity export and trading agreement

1. Overview

Storage projects require an agreement with an offtaker in order to purchase the electricity exported from the project onto the wider network. Given the efficiency losses of all storage technologies, an export agreement alone will more often than not be a cost, rather than a source of income, as the amount of electricity exported will never be as much as the electricity imported into the storage asset. Partly for this reason, there has been little demand for long-term route-to-market electricity export arrangements, unlike in the renewables sector. However, in regions where there is significant electricity price volatility and therefore price arbitrage is viable, the price received by the storage project for exporting in periods of high wholesale electricity prices could be significant.

Many of the terms of storage export agreements are familiar from other generation export arrangements, such as discount to market pricing and commissioning, outage and other information requirements. Nevertheless, a storage export agreement should reflect the revenue stack of the relevant storage project, such that when the project is required to export in order to provide a relevant service, the offtaker will purchase such electricity. Given that such services are frequently dispatched in effectively real time, with little forward visibility, this requires the offtaker to manage its imbalance position carefully.

The offtaker may offer dispatch of the project as a service; however, the parameters of this will depend on the relevant project, the equipment installed and the services that the storage is seeking to provide. Further, where a project opts for a combined import and export arrangement, it is possible for storage projects to outsource their state of charge management to the offtaker. This approach has the advantage of potentially optimising the trading and price arbitrage opportunities for the project. However, the interplay of the O&M agreement, the technology provider warranty, any aggregator involvement and the offtaker role adds complexity and could mean that there is more limited protection for the project where project issues arise that have or could have multiple causes.

2. Trading arrangements
In light of the shift in focus away from longer-term contracted revenue streams, the market access that the offtaker may be able to offer developers and investors is of increasing importance to storage projects. This can include forward, day ahead and intraday electricity markets, and balancing trading mechanisms. Where such access is required, the offtaker will require industry standard testing in order ensure that the storage project is capable of providing the capacity traded at the required technical requirements (eg, ramp rate, speed

"In light of the shift in focus away from longer-term contracted revenue streams, the market access that the offtaker may be able to offer developers and investors is of increasing importance to storage projects. This can include forward, day ahead and intraday electricity markets, and balancing trading mechanisms."

of discharge and profile). The project may also be required to install additional metering and/or equipment to allow for dispatch and monitoring of the project by the offtaker.

The pricing of the provision of such market access depends on the negotiated position between the parties and can including fixed fees for placing trades and/or profit sharing arrangements. Depending on the exposure that the offtaker is taking on the traded position of its counterparty, credit support from the storage project may be required. In addition, the project may be exposed to significant liability for imbalance position and any non-delivery charges where it fails to deliver the capacity traded.

Nevertheless, any trading arrangements ought to reflect any other services that the storage project has committed to provide that are incompatible with trading – for example, the commitment to be available for provision of an ancillary service may mean that the project cannot participate in trading in that settlement period.

3. Co-located projects

The export arrangements for co-located projects require particular consideration, given the interface with the generation element of such project. Various features of market-standard generation power purchase agreements may not be fit for purpose for a generation-plus-storage project, including:

- requirements to maximise the amount of electricity exported from the generating station (which may not be fulfilled given that storage technologies are not 100% efficient);
- provisions on forecasting and access to generation and export data, particularly in relation to solar projects, whose export profile may shift significantly;
- restrictions on providing ancillary services or participating in the balancing mechanism without the offtaker's consent;
- provisions requiring the revenue sharing of any 'new benefits' with the offtaker;
- metering provisions, particularly where additional metering is required to allow for the participation in specific revenue streams; and
- the possibility to trade the storage capacity in the wholesale market.

The commercial terms of the electricity offtake agreement for the co-located project will be influenced by the contractual structure – for example:

- the offtaker may seek to share in the combined project's storage

revenues (eg, ancillary services and balancing); and
- conversely, the generator may seek to negotiate a reduced discount to the relevant electricity index or market price as a result of being able to better manage the imbalance risk of any intermittent generation more effectively.

Many co-located schemes seek to have a bundled power purchase arrangement in which the offtaker agrees to purchase all power produced from the combined project. Alternatively, it may be possible to facilitate separate power offtake agreements, depending on the market, grid and metering arrangements.

IX. Import electricity supply agreement

1. Agreement structure

Energy storage projects require import electricity in order to function. Depending on the project structure, the legal arrangements to facilitate this can be structured by way of:

- an import agreement with a licensed electricity supplier, allowing for import of electricity from the wider network. This is typical for standalone projects;
- for behind-the-meter projects, an on-supply agreement between the storage project and the relevant demand customer. This is typical where the demand customer is registered at the import connection point for the site and therefore is responsible for the import of electricity at the site pursuant to its electricity supply agreement with a licensed electricity supplier. This arrangement is possible only where the on-supply of such electricity is allowed under the relevant jurisdiction's regulatory framework (eg, by providing that no electricity licence for such supply is required);
- for co-located projects, a supply agreement between the storage project and the co-located generation project. If the projects are part of the same corporate entity, this arrangement may not even be contractualised. Such arrangements may contribute to optimisation of use and value of the generated

electricity. Nevertheless, such on-site supply arrangements are rarely the only import supply arrangements for a storage project, particularly where the generation project is an intermittent technology. This is because the storage project will often wish to have flexibility over its state of charge management and to ensure that the storage project can:

- meet its contractual obligations to provide capacity and/or system services, and therefore maximise revenues and avoid any penalties for unavailability/non-performance; and
- be ready to capitalise on wholesale market fluctuations by importing during periods of low or even negative pricing and exporting during periods where the network requires additional supply.

These arrangements are often combined with the export agreement, which enables the state of charge management to be more effectively managed.

2. Pricing

The import of electricity represents an operational cost for storage projects. As a result, it is important that the import pricing is aligned with the revenue streams that the project is targeting to the extent possible, particularly where such revenue streams are paid on a delivery rather than availability basis.

In jurisdictions in which negative pricing has become an increasing feature of the electricity market, this offers an opportunity for storage projects to gain an additional revenue stream in those periods. However, given the wider issues that negative pricing can present to the wider electricity market, it is not expected that such circumstances will be a significant element of the revenue stack of such projects, given that governments and regulators are likely to seek to reform the regulatory framework to address such adverse consequences.

One perennial issue for storage developments is which elements of the wider electricity price, over and above the wholesale price, electricity storage projects should be charged in respect of such imported electricity – for example, should electricity storage be liable for environmental levies? This depends on the application of such levies in various jurisdictions. If this is charged on 'final consumption', but does not distinguish between whether most of the electricity will subsequently be re-exported to the system, this fundamentally disadvantages storage as compared with other electricity stakeholders. Nevertheless, no storage technology is 100% efficient and therefore there will always be a degree of consumption; as a result, there is a question as to whether other end users should cross-

"The import of electricity represents an operational cost for storage projects.
As a result, it is important that the import pricing is aligned with the revenue streams that the project is targeting to the extent possible, particularly where such revenue streams are paid on a delivery rather than availability basis."

subsidise storage technologies with lower efficiencies. However, many governments are conscious that storage is a fast-developing area and therefore are trying to legislate in a technology-neutral way (see Chapter V, section 5). Different countries have attempted to address this issue in different ways.

Case study: Great Britain
In Great Britain, Ofgem has effectively designated storage as a subset of generation. As a result, it is possible for projects to obtain a generation licence, which in turn exempts the licence holder from certain final consumption levies by virtue of the definition of 'supply' under the Electricity Act 1989:

'Supply', in relation to electricity, means its supply to premises in cases where:
(a) it is conveyed to the premises wholly or partly by means of a distribution system; or
(b) (without being so conveyed) it is supplied to the premises from a substation to which it has been conveyed by means of a transmission system,

but does not include its supply to premises occupied by a licence holder for the purpose of carrying on activities which he is authorised by his licence to carry on.[65]

However, this is not a perfect solution for all of the applicable final consumption levies in Great Britain. Therefore, some further regulatory amendments are underway in respect of the contract for difference and capacity market levies, and a separate notification remains under discussion in respect of the climate change levy. There are various other issues with this approach, such as the following:

- It requires smaller-scale storage projects to obtain a licence where equivalent other generation projects simply benefit from the applicable exemption regime; and
- There remains the wider fundamental question of whether storage is best defined as a subset of generation.

X. Grid connection arrangements

1. Key considerations

As with any generation project or consumption site, the grid connection arrangements for storage developments are critical. In particular, the process for obtaining the connection, the timing of the provision of such connection and the cost of such connection can make the difference between a viable project and a project that does not justify its initial capital investment.

The specific contractual structure in relation to the grid connection will vary by jurisdiction. However, universally, the approach to storage connection arrangements depends on the eligibility and suitability for the revenue streams that the project is currently targeting and may wish to provide in the future – the latter of which, as set above, are becoming increasingly important.

The considerations include:

- whether to connect to the high or low voltage network. This decision can affect the terms of the connection – in particular, the timing and cost of the connection;
- the upfront costs of the connection – for example, to pay for any specific upgrade works that are required to facilitate the connection;

"The grid connection arrangements for storage developments are critical. In particular, the process for obtaining the connection, the timing of the provision of such connection and the cost of such connection can make the difference between a viable project and a project that does not justify its initial capital investment."

- the use of system charges that may be payable, which in certain circumstances may be negative, resulting in additional revenue potential for the project;
- the firmness of the connection – that is, whether the project has the right to import and export on to the network at all times or whether it could be constrained off; and
- other technical requirements, such the allowed ramp rate (ie, the speed at which the storage project can change its output) and the power swing (ie, the potential change in export to import and vice versa).

Given the characteristics of storage projects, a range of technical reasons can mean that a particular grid connection location is unsuitable – for example, constraints on either import or export capacity or ramp rates. Nevertheless, standalone grid connection applications for proposed storage projects have proliferated in certain jurisdictions – for example, in early 2017 one DNO in the United Kingdom reported that it had received 12.2GW of applications for storage projects over 15 months.[66] While many of these speculative applications have not resulted in delivered projects, this surge in interest in grid connections has led to severe constraints in the pipeline of projects and resulted in various delays in processing such applications. Further, network owners are also incentivised to be conservative by the regulatory framework (eg, licensing and wider legislation and regulations). This has led to accusations that storage is considered in the worst-case scenario – for example, treating storage as intermittent and operating to exacerbate system issues during times of system stress – in order to ensure the security of the network. This is opposed to recognising the wider benefits that standalone projects can provide to the network and the high likelihood that storage projects will be incentivised to operate in a way that minimises system issues.

2. Existing connections

To date, the use of existing connections for co-located renewables projects and behind-the-meter storage projects has been preferred, as it optimises the existing connection, which has already been secured and delivered, and improves the utilisation of any renewable electricity produced on-site. This approach also reduces project costs and the risk of project delays. However, this may still require the amendment of the terms of the existing connection agreement – for example, an increase in import capacity to enable the storage device to charge from the system. In any event, it is likely that the network counterparty to the connection agreement will need to be informed. Further, interface and sharing arrangements (eg, dealing with costs and maintenance responsibilities, and potential liabilities) between the generation project or consumer and the storage project may be required to regulate utilisation of the connection.

3. Active network management

Active network management is a form of flexible connection that has been particularly associated with storage. This type of connection allows the network operator to monitor and control the operation of generation and demand by installing software at the relevant sites, which can include limiting the ability to utilise the network at particular periods. Active network management can avoid reinforcement of the network and therefore reduce the costs and time to connect.

The simplest order of priority of connection is a 'last in, first out' approach; however, the flexibility of storage means that this is not necessarily the most efficient way to manage the network. For example, storage can assist in reducing peak demand or providing additional capacity to avoid the need for other generation or demand to be constrained. However, the treatment of storage has been inconsistent – in particular, whether the project is paid for providing such a service to the local network.

"Many of the issues to be considered in the construction of storage projects are principally the same as those in the wider clean energy sector. For example, it is usual for the contractor to be liable for liquidated damages for delay."

XI. Construction contract

Given the relative immaturity of the storage sector, there is yet to be a standard approach to the structuring of construction agreements for storage projects. This contrasts with other technologies such as solar, where a combined EPC agreement is usual; or onshore wind, where the turbine supply and maintenance arrangements tend to be separated from other contracts.

1. Contractual structure

As with other clean technologies, where a wrapped EPC approach is adopted in respect of storage projects, the additional risk placed on the EPC contractor can result in a costlier agreement. In jurisdictions where storage sector revenue streams are under pressure, this has resulted in developers exploring alternative multi-contracting approaches – for example, providing for a separate supply contract for the storage technology and a balance of plant contract. Where a multi-contracting approach is adopted, the scope of the balance of plant agreement may cover the civils and electrical installation and the supply of other balance of plant equipment such as switchgear. The aim of this approach is to achieve a lower construction cost in return for the developer accepting greater interface risk between the various contracts. Nevertheless, this could also result in significantly reduced protection for the project (eg, splitting of liability caps), if not designed carefully. Issues to be considered in relation to a multi-contracting approach include the following:

- The potential interface risk is clearly increased where a multi-contracting approach is adopted – for example, delay to the construction programme which may have multiple causes. Contractors are often reluctant take the risk of third-party interference if a delay to completion would result in liquidated damages being payable. As a result, the project's delay protection is likely to be reduced as the balance of plant contractor will not take on the risk of the technology provider's delay and vice versa; therefore, this may require the developer to pursue a number of contractors. In addition, contractors are likely to seek to allocate this risk to the employer by requesting that an act of prevention by another contractor give rise to an extension of time and additional costs.
- In addition, consideration should be given in particular to the physical interfaces between the various contracts, and in particular who would be responsible should physical damage arise. Other issues include the impact of construction and commissioning tests of the various works and the responsibility for risk of defects on the performance of the assets.
- The warranty provided by the balance of plant contractor is likely to be more limited – for example, only covering installation of the storage device. Further, it is possible that the balance of plant contractor may be a less creditworthy entity, which may result in the need for enhanced credit support to stand behind any warranties provided or the project developer accepting greater insolvency risk.
- Further, a multi-contracting approach requires more in-depth consideration of the battery waste recycling issues discussed in Chapter XIV, as there is a greater risk that the developer could be deemed to be the 'producer' in a multi-contracting situation.

2. Key issues

The approach to the construction arrangements will be influenced by the specific storage technology selected for the project. As an example, the construction of a pumped hydro project requires a very different contractor skillset and has a significantly different delivery risk profile as compared to the supply and installation of a containerised battery storage solution.

However, many of the issues to be considered in the construction of storage projects are principally the same as those in the wider clean energy sector. For example, it is usual for the contractor to be liable for liquidated damages for delay.

Of particular focus for energy storage construction agreements are the following:

- the form of performance guarantees provided by the contractor – for example, whether the guarantee is based on capacity, availability or both. The alignment of the performance guarantee to the revenue strategy to be pursued should also be considered. The quality of the guarantee of performance will be a key area of focus for contractor selection, particularly given the nascent stage of development of the storage sector;

- the technology provider's warranty, which should be long term and comprehensive. On the scope of the warranty, this may be closely aligned to the current business plan or provide a degree of latitude to encompass a change in revenue strategy. The latter approach may mitigate against any future risks of changes of income flows or operational approach, but may mean that there are gaps in the coverage initially that are passed on to the employer;

- where the technology provider's warranty is held by the contractor, provision should be made for its assignment to the employer in specified circumstances, such as where the caps on liability have been reached or the EPC contract has been terminated. Further, the EPC contract should provide for assignment of this warranty to lenders by way of security in the case of debt financing;

"There are additional issues to consider when negotiating construction contracts for storage projects for non-standalone applications, such as co-location and behind-the-meter, ranging from access rights to shared infrastructure."

- the experience and credibility of the contractor and/or the technology provider, in particular for more novel technologies. This extends to their creditworthiness – specifically, their ability to stand behind the performance guarantee and/or warranty. Where this is not deemed sufficient, performance security by way of guarantees, bonds, letters of credit or otherwise should be obtained to provide the employer with the level of comfort required;
- the level of damages payable by the contractor in the event of delay and/or if the performance guarantee is not achieved. These arrangements come under particular scrutiny where a contracted revenue stream requires the project to be commissioned by a specific date and/or prescribe specific penalties. Such damages are likely to be set as liquidated damages and subject to caps on liability, which are usually set a percentage of the relevant contract price;
- the testing and acceptance regime – in particular, whether the regime aligns with the performance guarantee, technology provider warranty and/or specified revenue streams. At the very least, the minimum level of testing should meet or exceed the key initial revenue stream contract in order to ensure that the employer is in a position to seek recourse against the contractor for lost revenue. However, as it becomes increasingly apparent that the business models of energy storage projects require a degree of flexibility in order to react to the evolving market and value chain, developers may move away from hardwiring in specific testing requirements; and
- the right for the employer either to reject or terminate the agreement in specified circumstances – for example, for extended delay or where liability caps have been reached. The position in relation to this and the consequences of such rejection and termination tend to vary significantly, depending on the negotiated position between the parties.

There are additional issues to consider when negotiating construction contracts for storage projects for non-standalone applications, such as co-location and behind-the-meter, ranging from access rights to shared infrastructure. However, these tend to be application and site specific.

XII. Operations agreement

1. Structure and key provisions

Once the construction risks have been successfully navigated, the operational management of the storage project comes into focus. As alluded to elsewhere in this report, these arrangements are of paramount importance for storage projects as compared to other generation projects, in order to optimise the project's availability, ensure access to and maximise the required revenue stream stack, and seek to prevent the devaluation of the storage technology itself by way of degradation.

The storage O&M agreement should:

- set out the specific operational parameters to which the performance guarantee from the EPC contractor and the technology provider warranty are likely to be subject – for example, the state of charge, number of cycles and depth of discharge;
- provide for preventative monitoring and maintenance of the storage project in line with the technology provider's warranty, such as monitoring of the performance data and physical conditions, such as battery cell temperatures, and physical inspections of the asset itself;
- address any serial and latent defects in the storage technology

and/or operational underperformance of the storage asset as and when these arise;

- where relevant, address degradation of the storage asset, providing upgrades of any required operational software and, if negotiated, providing any requested capacity extensions; and
- provide for access to the storage project for the counterparty to any revenue stream contract, such as the relevant system operator, aggregator or offtaker, in order to allow for independent testing or inspections to be carried out in accordance with the terms and conditions of the relevant revenue stream contract.

2. Additional considerations

The duration of the O&M agreement can vary considerably, depending on the negotiated terms, the term of the contracted revenue stack, the term of the technology provider warranty and the risk appetite of the developer, its investors and lenders. Further, the structure of the operational arrangements will be influenced by the approach taken for the construction agreements – in particular, whether a wrapped or multi-contract approach has been taken. If the storage assets have been supplied by the technology provider to the employer, the maintenance of the storage asset may be undertaken directly by that entity, with a supplementary balance of plant maintenance agreement with a third party put in place. If a wrapped EPC contract approach has been taken, then it is more likely that a single O&M agreement will be signed with the same entity.

The interplay between the O&M agreement and the export, import and dispatch arrangements should be carefully considered. In particular, who bears the risk of interference by a third-party entity on availability and performance is an issue that will require careful consideration by both technical and legal advisers. A range of operational agreements may be put place in addition to the O&M agreement, depending on the desired risk allocation, such as management services and asset management agreements for the provision of company administration and financial services, and in-house asset management of the storage asset.

There will be additional considerations for O&M arrangements for co-located and behind-the-meter projects, including whether any optimisation or cost savings can be achieved through shared site security, monitoring, maintenance and asset management. Naturally, additional interface and access considerations come up in co-located and behind-the-meter projects.

XIII. Land agreement

1. Land requirement

The amount and location of the land required for a storage project can vary significantly, depending on the choice of technology (eg, a pumped storage project requires a large amount of land in very specific geographical locations) and revenue streams being targeted by the project (eg, grid connection availability in the relevant area or the need to participate in a local flexibility market which has locational requirements). Further, it is important for developers and funders to ensure that sufficient land rights are available in order to connect the development to the grid, so as to prevent project delay and ransom strip issues.

The timing of securing the land rights can be dictated by the timetable of the relevant procurements for revenue streams. For example, it is not uncommon for an auction to require that a project has secured at least an option for the relevant site before being eligible to participate in the relevant auction. In other circumstances, the relevant construction lead-in times can mean that land is secured in shorter order – for example, battery projects can be constructed in a matter of weeks and therefore the land rights are not required significantly in advance of the intended operation date.

2. Key considerations

Where the project does not own the land outright, the term of the relevant land agreement will be pertinent. Developers are seeking to learn lessons from more mature technologies – for example, wind, where the asset life has been significantly extended and further asset extension methodologies are being developed throughout the project lifetime – by allowing for longer terms than may be currently anticipated for the relevant technology. Nevertheless, there is a balance to be struck between the flexibility for life extensions in future years and the upfront commercial deal that the developer is seeking, particularly in jurisdictions where there is relative revenue uncertainty.

Rental arrangements can vary significantly; however, it is not uncommon to see rent based on per MW of initial installed storage capacity. Maximum capacity limits within the permitted use provision can cause issues – such as preventing the implementation of technology improvements, future capacity extensions or asset optimisation – during the life of the project. In other generation projects, land owners have often tried to link rents to a proportion of the revenues, driven by certain lucrative subsidies. Such an approach is not suitable for the majority of storage projects, given the revenue stacking approach and revenue stream flexibility required in most jurisdictions.

"The utilisation of energy storage behind-the-meter or co-located with generation projects throws up some further interesting issues in relation to land rights – for example, where the storage system sits within the site and whether the current permitted and future use sections of any land rights arrangements allow for the installation and operation of storage."

Depending on the project structure and funding required, the transferability of land rights may be an important consideration, particularly where asset sales to investors and/or debt financings requiring restructuring are envisaged. As with other projects of a similar nature, the land rights agreements must pass the bankability requirements of lenders and allow for step-in rights for lenders. End-of-life issues should also be considered during the land rights negotiation stage (eg, the decommissioning and restitution of land requirements), particularly where significant engineering work has been required for the relevant technology.

The utilisation of energy storage behind-the-meter or co-located with generation projects throws up some further interesting issues in relation to land rights – for example, where the storage system sits within the site and whether the current permitted and future use sections of any land rights arrangements allow for the installation and operation of storage.

"The most significant potential cost placed on producers under the Batteries Directive is the so-called 'take-back' obligation, which requires producers to take back the storage device from users at the end of its life free of charge."

XIV. Permitting

1. Consent to construct and operate

How storage projects sit within the planning regime depends on the jurisdictional legislation and regulatory requirements, which very often were not designed with energy storage in mind.

In many circumstances, the total project capacity is a key determining factor for which planning routes are available. For example, in Great Britain, if the project is over 50MW, it will fall within the development consent regime, which dictates a specific process that is costlier and more time consuming. As with land rights, having obtained the key planning permits is often an eligibility requirement in order to participate in certain revenue stream procurements.

A variety of options are available to a developer, which can include:

- applying for a standalone planning permission. Irrespective of jurisdiction, it is likely that larger-scale storage projects will require separate planning permits in order to construct the project;
- varying an existing planning permission, particularly for co-located or behind-the-meter developments; and
- utilising existing permissions available to the development – for example, under any electricity licence.

Particularly for newer technologies, some planning authorities may require additional assistance in order to understand the project and the potential risks involved. Planning conditions are likely to be technology specific, but can include noise during the operational phase and mitigants to deal with any explosion or fire risk.

In addition to planning consent, a storage development may need additional licences to operate. For example, in Great Britain, if the project is over 100MW, it will require a generation licence, as in Great Britain storage has been classed as a subset of generation. Projects below 100MW may choose to obtain a generation licence, which has the benefit of exempting the project from final consumption levies on electricity imported to the storage device from the network; or alternatively may benefit from a specific or class exemption from the requirement to hold a licence. This is also a relevant consideration for co-located projects in terms of the total generating capacity of the combined project; as a result, the licensing status of the entire development may need to change.

2. Producer responsibility

The nature of chemical energy storage technologies can result in additional environmental considerations. For example, the EU Batteries Directive (2006/66/EC) places certain obligations on 'producers' of batteries and covers larger-scale energy storage systems.

The most significant potential cost placed on producers under the Batteries Directive is the so-called 'take-back' obligation, which requires producers to take back the storage device from users at the end of its life free of charge. Breach of the Batteries Directive implementing regulations can be a criminal offence; as a result, it is important that all parties understand who is taking on such liability.

The definition of 'producer' under the Batteries Directive is wide, focusing on the entities that place the batteries on the relevant market, and can include manufacturers, suppliers, importers, developers and EPC contractors, depending on the implementation of the Batteries Directive in the relevant member state and the applicable contractual structure. As a result, developers are often keen to structure their contracts such that they are not considered to be a 'producer' under the national legislation implementing the Batteries Directive.

3. Wider environmental and health and safety requirements

As emphasised above, each project should be assessed on an individual basis, depending on the technology in the context of the relevant national regulatory regime. In addition to the relevant planning consent required, the following themes are common:

- Wider hazardous waste requirements may apply to chemical storage technologies in addition to the producer responsibility requirements set out above, particularly if there is any leakage;
- Environmental permits may be required in addition to any planning consent, particularly where there are any emissions into the environment. Storage systems that utilise water, such as pumped hydro, may have to comply with various water abstraction requirements;
- Packaging, labelling and transportation requirements are highly technology dependent – for example, in relation to whether they are caught by the European Agreement concerning the International Carriage of Dangerous Goods by Road; and
- Although health and safety law in the vast majority of jurisdictions does not specifically contemplate energy storage, such legislation will nonetheless apply to such projects.

The relevant regulators in various jurisdictions are considering guidance for storage developers on these issues in order to provide clarity to the developing storage industry.

*"Company structures should be mindful
of any relevant unbundling requirements,
ensuring that the required structural
separation is achieved, and that confidential
information and personnel are dealt with
in a compliant manner."*

XV. Corporate arrangements

1. Corporate structures

As has been illustrated above with other legal elements, there are no typical corporate arrangements for energy storage projects. However, the use of project-specific SPVs that hold the project rights and assets for a single storage project and asset companies holding multiple energy storage projects is common.

However, the complexity of the corporate structures of energy storage groups is anticipated to develop as:

- debt financing activity increases and debt and security structures become more complex (eg, portfolio and cross-collateralised structures); and
- further equity investment at a group level is attracted into the larger developers and technology providers.

Such developments are likely to include:

- the insertion of various holding companies, to facilitate lenders taking security over a tranche of projects and/or to allow preferential access to ancillary and balancing markets for the combined capacity of the underlying projects; and
- the creation of asset management and other operations

companies, such as trading companies, within the corporate group, which then contract intra-group to provide services to the relevant asset companies, where such activities are performed or brought in-house.

The above assumes that storage projects are the sole business line of the relevant company. However, this is not usually the case – for example, a number of utilities and infrastructure or other investment funds have invested in or developed storage projects, but also have a wide range of interests in the power sector. In these circumstances, company structures should be mindful of any relevant unbundling requirements, ensuring that the required structural separation is achieved, and that confidential information and personnel are dealt with in a compliant manner.

2. M&A activity

In terms of M&A activity, there has been significant activity for individual projects during the development window (ie, usually once the relevant project has secured its relevant planning, land and connection rights). This has typically resulted from the developer not having the capital to build out the project and investors not having the appetite for early-stage development risk, mirroring a trend in the wider renewables

"Development agreement structures have also been explored, allowing investors to have exclusivity of a portfolio of in-development projects and purchase majority stakes or the entirety of individual projects once they have met pre-agreed milestones."

market. These deals have involved share and asset sales depending on the seller's corporate structure. In different jurisdictions, there have been a variety of approaches to the extent that contracted revenues are required to be secured already in order to attract incoming investors.

Development agreement structures have also been explored, allowing investors to have exclusivity of a portfolio of in-development projects and purchase majority stakes or the entirety of individual projects once they have met pre-agreed milestones. This allows investors the ability to secure a pipeline of projects at an early stage, without the associated early-stage development risks. Nevertheless, such structures are often complex to negotiate and implement, as there is a need to anticipate all the various permutations that may arise in a project's development cycle and provide for the required amount of certainty and discretion for the incoming investor.

The jurisdictions that have had the most significant energy storage activity have also been the key jurisdictions for M&A activity. According to BDO analysis of energy storage deals between 2010 and 2018, the acquisition target's country of origin was predominantly the United States (52% of deals), followed by Germany (16%), France (8%) and the United Kingdom (8%).[67]

3. Investors

Various traditional power utilities have set up energy storage ventures. Some, such as Vattenfall and EDF, have developed their own storage businesses in-house by participating in various auctions in relevant jurisdictions. EDF aims to achieve 100% carbon-free power by 2050, with energy storage playing a significant role in achieving this goal: 5 GW of installed grid-scale storage is already available and the company plans to increase that to 15 GW by 2035.[68] However, other utilities have bought into:

- development pipelines, such as Statkraft's acquisition of the storage arm of Element Power, alongside its significant wind pipeline in October 2018;[69] and
- individual projects, such as Enel's acquisition of the 25MW Tynemouth battery storage project in the UK.[70]

Wider equity investment in energy storage has resulted in incoming investors investing in entire storage groups. For example:

- on the developer side, in February 2017 a joint venture of Low Carbon and VPI Immingham created VLC Energy to fund early-stage energy storage and renewable energy projects, including the multimillion-pound development of two energy storage projects, in the United Kingdom;[71] and

- on the technology side, AES and Siemens merged their storage subsidiaries to form a 50:50 joint venture named Fluence, which started trading in January 2018. Fluence has become a key contractor stakeholder for the development of battery projects in a number of jurisdictions.[72]

Depending on the structure of such investments, these transactions have required joint venture shareholder agreements between the parties, the terms of which are highly dependent on the commercial terms of the relevant deal.

Case study: strategic investment
In July 2017 temporary power provider Aggreko bought Younicos, a developer of battery technology and energy storage solutions, for £40 million. The deal enables Aggreko to offer a range of storage solutions to its customers as part of its wider flexibility offering.[73]

Case study: residential behind-the-meter investment
In May 2018, Shell participated in a €60 million investment round by German residential battery storage firm sonnen.[74] In February 2019 it was announced that Shell would acquire 100% of sonnen for an undisclosed sum, with the stated aim that the acquisition will allow Shell "to offer more choice to customers seeking reliable, affordable and cleaner energy".[75]

Dedicated energy storage funds have also been used as a means of investing in energy storage, most notably by SUSI, which raised €252 million from institutional investors in May 2018.[76] The use of listed investment funds may also increase. There have been two initial public offerings to date at the time of writing, both on the London Stock Exchange:

- Gore Street Energy Storage Fund listed in May 2018, raising circa £30 million; and[77]
- Gresham House Energy Storage Fund listed in November 2018, raising circa £100 million.[78]

However, both funds failed to raise what was originally targeted.

XVI. Financing arrangements

1. Overview

Debt financing of energy storage assets – particularly of emerging technologies – is still a developing area. However, there have been examples of increased appetite from debt providers to finance such assets, as they become more comfortable with the relevant energy storage technology, the tendency to utilise a revenue stream stacking approach and the degree of merchant risk associated with the relevant energy storage application in the relevant jurisdiction.

Much of the debt financing deployed to date has related to specific projects, even if structured such that the debt is provided higher up the corporate group. It is anticipated that in the future there are likely to be more group financings, as adopting a portfolio approach allows the spreading of risk across a variety of assets of different types and the increased scale will enable project risk to be spread across a cross-collateralised group.

Further, the timing of securing debt financing has varied, as the construction risk varies significantly depending on the storage technology. For example, financing or re-financing pumped hydro following the construction period significantly reduces the risk profile of the project, when compared to a containerised battery solution.

2. Key challenges

At present, while it is clear that there is appetite from lenders to finance storage projects, particularly certain battery storage projects, there are not yet repeat project financings or significant scale of financings in the storage market for a variety of reasons, including the following:

- The nature of the developer companies is such that those seeking project financing are newly established companies with little track record, which makes each financing more challenging to deliver;
- Not all lenders are currently able or interested in lending to storage projects at present. While certain lenders have capital to deploy in this area, they are keen to ensure that they are not overexposed to any one sector, so doing one or two landmark and/or pathfinder transactions (no matter what the size and on what terms) may seem to be sufficient in light of the current market;
- There is a limited number of good-quality projects available to finance – in particular, those with a lower proportion of income from merchant revenue and/or a higher proportion of longer-term contracted revenues;
- It is difficult to predict the future value of flexibility, meaning that lenders are reluctant to back a significant number of projects in any one jurisdiction;
- The lack of a repeatable financing model is also a barrier, as a number of developers have different and evolving business models, and are raising debt at different times for different reasons with different structures;
- The cost and higher margins of the debt financing can still be relatively expensive and this may be limiting the development of the market; and
- The scale of the storage market is a brake on the deployment of energy storage financings, as the ticket sizes currently remain small, which will limit the attractiveness to some lenders. As a result, there is a concern that the sector will remain a niche area from a financing perspective.

3. Key issues

As is evident from this report, there is a huge variety of business models in energy storage, most of which are more complex than those deployed in other segments of the power sector, such as renewables. Lenders need to understand the business model currently proposed, including the structure of the revenues (eg, whether these are based on availability or delivery payments), the level of revenue certainty, the possible penalties that could apply and other risks of that revenue not being available. In models that involve a third party – for example, behind-the-meter applications that rely on savings accruing to a corporate entity – this will also involve an assessment of that counterparty and its creditworthiness, particularly as such arrangements tend to be long term. This process can be time consuming, particularly as future regulatory and market changes will also need to be allowed for.

There are a number of issues that debt providers will focus on when structuring a debt financing package for a storage project:

- Revenue certainty – in particular, the term of the relevant revenue stream and how it is accessed – is a key consideration. Lenders may well value greater certainty with a lower return for a proportion of revenues over more volatile and unpredictable revenues that may have a higher return. Nevertheless, as the market in storage evolves, it is likely that lenders will have to accept at least a degree of merchant revenue exposure.
- Lenders will want to understand the creditworthiness of the counterparties to the various revenue streams. In most cases where the revenues are coming from network and/or system operators, the licence requirements and wider regulatory regime are available to provide lenders with the necessary comfort. However, solvency concerns do arise, as illustrated by the issues currently faced by PG&E.[79] Creditworthiness can be an issue

"While developers of storage projects are closer to the regulatory landscape and how it is evolving, it is often not in their control; however, the view is that they are better placed to manage this risk than any debt providers to the project."

where developers use an aggregator to access such markets, as the lender is exposed to the potential insolvency of the aggregator, which is often a thinly capitalised company with a limited trading history. This is exacerbated where the relevant revenue stream provider does not provide any additional comfort to the end storage provider, such as step-in rights. As more innovative business models are developed, such as the use of flexibility platforms, this may further complicate the picture.

- Revenue stacking can diversify risk and ensure that an asset is not overly exposed to any one revenue stream. However, such an approach adds complexity and the risk that the competing demands on the project may result in failure to deliver or damage to the asset. As a result, where a project is following this strategy, lenders will want to understand the combination of revenue streams and their interaction with the storage technology and the warranty and O&M package available.

- Regulatory risk in the storage sector remains a significant barrier. Announcements such as the Tempus state aid decision in relation to the Great Britain capacity market in November 2018, which resulted in the suspension of payments under existing agreements, have also served to undermine lender confidence in longer-term contracted revenues.[80] This has led to more conservative debt terms such as lower loan to value ratios and shorter debt repayment terms. While developers of storage projects are closer to the regulatory landscape and how it is evolving, it is often not in their control; however, the view is that they are better placed to manage this risk than any debt providers to the project.

- Technology risk and performance is an area of focus for lenders, particularly for newer technologies, where few or even no projects have reached their projected lifetime. However, lenders have demonstrated that they can get comfortable with assets such as lithium ion batteries despite inherent technology risks such as degradation. They have been able to do so through a combination of greater technical understanding, confidence in the business plan, clear operational parameters and good-quality warranties from a creditworthy battery storage technology provider. A reputable technology provider can be an important factor. This is a barrier to entry to the debt market for newer technology providers and projects, which tend to be smaller, with less of a proven track record.

- Debt providers are likely to prefer a full-wrapped EPC contract, rather than the multi-contracting approach described above, as the former should de-risk the project further from their perspective. This may be at odds with the developer and equity investors, which may be more comfortable taking additional interface risk in return for a lower capex construction cost.

"The operational aspects of an energy storage project are crucial to its long-term success, particularly for certain technologies such as batteries, which can be adversely affected when not operated correctly. As a result, the operational arrangements such as asset management will be heavily scrutinised by lenders."

- As highlighted above, the operational aspects of an energy storage project are crucial to its long-term success, particularly for certain technologies such as batteries, which can be adversely affected when not operated correctly. As a result, the operational arrangements such as asset management will be heavily scrutinised by lenders. Operating systems that can provide additional certainty about accessing revenue streams in near-to or at real time may also be attractive.
- As with other emerging sectors in the energy sector where the risks cannot be mitigated contractually, lenders will expect the developer to take any increased risk when compared with the traditional non-recourse project financing of subsidy-backed renewables projects or conventional power projects.

One area that is attracting interest from debt providers is the co-location subsector of storage projects. This business model can allow for diversification of risk between the storage and generation elements. Indeed, in some circumstances, the lenders may simply seek to understand the generation project and then, on the basis that the revenues for that asset alone are sufficient to cover the debt, lend against a project. In addition, the storage aspect of such projects is often a fairly small proportion of the overall capacity at present – for example, the Lincoln Gap project in Australia is a 212MW wind farm with a 10MW battery co-located, which secured A$150 million from the Clean Energy Finance Corporation.[81] Nevertheless, the financing of such co-located projects can be more complex – for example, to the extent that the generation element has any existing debt financing or a hybrid corporate ownership structure in place.

4. Key financing terms

Nevertheless, despite the key issues and challenges highlighted above, a leading credit agency recently stated that it finds the "financing approach of a battery storage project to be broadly akin to many of the risks associated with financing a conventional power project".[82] There have been a range of debt financings across the world, such as the following.

- A Macquarie Capital-backed battery-based energy storage project located in Southern California secured $100 million of debt in December 2018 and a further $35 million in March 2019. The backers were CIT Group, Rabobank, Sumitomo Mitsubishi Banking Corporation and ING. The financing is part of a 63 MW/340 MWh project across 90 different sites for 28 different host customers.[83]
- In early 2018 Santander lent £28.5 million to Battery Energy Storage Solutions, now known as Zenobe Energy. This was one of the first project finance transactions for battery storage in the United Kingdom.[84]

To the extent that a storage developer is looking to finance an energy storage project by way of debt, the terms of such financing would, among other things, address the following:

- **Equity contribution/loan to value ratio:** The debt providers are unlikely to finance the entire cost of the storage project, so an equity contribution will be required by the developer group. The amount of such equity contribution will be influenced by the debt provider's loan to value requirements. The determination of the required equity contribution will depend on the specific project, the sponsors and the individual credit requirements of the debt provider; however, the equity contribution has been substantial when compared to financings in the wider clean energy sector.
- **Length of debt term:** Long-term debt financing is unlikely, due to the short-term nature of the key revenue-producing contracts underpinning the storage project. Certainly, the terms seen to date are shorter than may be expected on a renewables project with a long-term subsidy support mechanism that a lender is more comfortable with.
- **Security:** The debt providers are likely to require a comprehensive security package, including debentures, legal

charges, shares charges, direct agreements, duty of care deeds and assignments in security over the relevant project documentation (including, but not limited to, any grid sharing agreements and contracts with aggregators). The registration of such security will depend on the jurisdiction.

- **Intercreditor arrangements/restricting other debt:** An intercreditor deed is likely to be required, to ensure that the debt providers have first priority of security and repayments ahead of the equity providers and any other junior creditors to the project until the senior debt is repaid in full. The debt providers are also likely to require a complete restriction on the borrower incurring any other debt for the duration of the loan, other than in certain limited circumstances where such debt or liabilities are fully subordinated.

- **Distribution lock-up:** The requirements of debt providers with regard to any distribution lock-up (ie, the ability for the borrower to make distributions to its shareholders/equity investors by way of dividend or repayment of any subordinated debt) will vary for each transaction, but these could range from a full distribution lock-up for the term of the loan to temporary restrictions that may, for example, apply where an event of default has occurred and is continuing, or where certain pre-agreed financial covenant levels have not been maintained. It may be possible for such restrictions to be lifted where the event of default is capable of remedy and is remedied within any grace period permitted by the debt providers or where the required financial covenant levels are restored.

"There remain a number of commercial, technical and legal challenges for the storage industry to overcome. However, such hurdles are not expected to halt the emergence of storage as a significant player in the electricity industry."

XVII. Conclusion

It seems inevitable that global large-scale deployment of storage will occur, due to the valuable flexibility services that such technologies can provide. However, the true extent of the opportunity for storage is not yet clear, in no small part due to the significant influence that the ever-changing needs of the wider electricity market have on the business case for such projects.

As illustrated in this report, there remain a number of commercial, technical and legal challenges for the storage industry to overcome. However, such hurdles are not expected to halt the emergence of storage as a significant player in the electricity industry. Over the coming years, further innovation in both the regulatory arrangements and contractual documents is anticipated, as this nascent sector matures and new business propositions, such as vehicle-to-grid and hydrogen storage, emerge.

Notes

1 www.iea.org/tcep/energyintegration/energystorage/#progress.
2 Bloomberg New Energy Finance Long-Term Energy Storage Outlook 2018,
https://about.bnef.com/blog/energy-storage-1-2-trillion-investment-opportunity-2040.
3 Bloomberg New Energy Finance Long-Term Energy Storage Outlook 2018,
https://about.bnef.com/blog/energy-storage-1-2-trillion-investment-opportunity-2040.
4 International Renewable Energy Agency (IRENA), Electricity Storage and Renewables: Costs and
Markets 2030, October 2017, p11, www.irena.org/-
/media/Files/IRENA/Agency/Publication/2017/Oct/IRENA_Electricity_Storage_Costs_2017.pdf.
5 IRENA, Electricity Storage and Renewables: Costs and Markets 2030, October 2017, p14;
www.irena.org/-/media/Files/IRENA/Agency/Publication/2017/Oct/IRENA_Electricity_Storage_
Costs_2017.pdf.
6 IEC Electrical Energy Storage White Paper; www.iec.ch/whitepaper/pdf/iecWP-energystorage-LR-
en.pdf.
7 IRENA, Electricity Storage and Renewables: Costs and Markets 2030, October 2017, p18;
www.irena.org/-/media/Files/IRENA/Agency/Publication/2017/Oct/IRENA_Electricity_
Storage_Costs_2017.pdf.
8 www.dnvgl.com/publications/2018-battery-performance-scorecard-132103.
9 www.greentechmedia.com/articles/read/green-battery-revolution-powering-social-and-
environmental-risks#gs.VgVGCgUB.
10 www.iea.org/tcep/energyintegration/energystorage/.
11 IEC Electrical Energy Storage White Paper, www.iec.ch/whitepaper/pdf/iecWP-energystorage-LR-
en.pdf.
12 Located at Huntor, Germany in 1978 and McIntosh, Alabama, United States in 1991.
13 www.irena.org/-/media/Files/IRENA/Agency/Publication/2018/Sep/IRENA_Hydrogen_from_
renewable_power_2018.pdf.
14 www.itm-power.com/sectors/power-to-gas-energy-storage.
15 IRENA, Electricity Storage and Renewables: Costs and Markets 2030, October 2017,
www.irena.org/-/media/Files/IRENA/Agency/Publication/2017/Oct/IRENA_Electricity_
Storage_Costs_2017.pdf.
16 IEC Electrical Energy Storage White Paper, www.iec.ch/whitepaper/pdf/iecWP-energystorage-
LR-en.pdf.
17 European Association for Storage of Energy, Energy Storage Technology Descriptions.
18 Panagiotis Bakos, "Life Cycle Cost Analysis (LCCA) for Utility-Scale Energy Storage Systems",
Journal of Undergraduate Research 9, 52-56 (2016).
19 IRENA, Electricity Storage and Renewables: Costs and Markets 2030, October 2017,
www.irena.org/-/media/Files/IRENA/Agency/Publication/2017/Oct/IRENA_Electricity_
Storage_Costs_2017.pdf.
20 https://anesco.co.uk/subsidy-free-solar-pv/.
21 www.ofgem.gov.uk/publications-and-updates/guidance-generators-co-location-electricity-
storage-facilities-renewable-generation-supported-under-renewables-obligation-or-feed-tariff-
schemes-version-1.
22 www.brooklyn.energy/ and www.businessgreen.com/bg/news/3030252/verv-completes-uks-
first-blockchain-green-energy-trade.
23 www.solarwirtschaft.de/index.php?id=875.
24 www.energy-storage.news/news/ireland-to-incentivise-solar-plus-storage-with-household-grants.
25 www.energy-storage.news/news/french-island-tenders-push-down-solar-plus-storage-prices-
by-40.
26 https://assets.publishing.service.gov.uk/government/uploads/system/uploads/
attachment_data/file/576367/Smart_Flexibility_Energy_-_Call_for_Evidence1.pdf and
https://assets.publishing.service.gov.uk/government/uploads/system/uploads/attachment_data/file
/633442/upgrading-our-energy-system-july-2017.pdf.
27 Clarifying the regulatory framework for electricity storage: licensing, 29 September 2017,
www.ofgem.gov.uk/system/files/docs/2017/10/electricity_storage_licence_consultation_final.pdf.
28 Electricity Generation Licence: Standard Conditions, www.ofgem.gov.uk/system/files/docs/2017/
10/elecgen_slcs_consolidated_29sept2017.pdf.
29 www.ofgem.gov.uk/system/files/docs/2018/10/smart_systems_and_flexibility_plan_
progress_update.pdf.
30 www.ofgem.gov.uk/system/files/docs/2018/10/smart_systems_and_flexibility_plan_
progress_update.pdf.
31 www.ofgem.gov.uk/system/files/docs/2019/01/storage_and_charging_reform_2201f.pdf.
32 This is not to say that there are not specific definitions in the underlying industry codes and
revenue streams arrangements. For further information, see Chapter V.
33 https://ferc.gov/whats-new/comm-meet/2018/021518/E-1.pdf.
34 https://eur-lex.europa.eu/resource.html?uri=cellar:c7e47f46-faa4-11e6-8a35-
01aa75ed71a1.0014.02/DOC_1&format=PDF.
35 www.sem-o.com/glossary/.
36 https://assets.publishing.service.gov.uk/government/uploads/system/uploads/
attachment_data/file/631656/smart-energy-systems-summaries-responses.pdf.

37 S Maynard and A Ason, "Is the Energy Charter Treaty Ready to Embrace Energy Transition?" (TDM, ISSN 1875-4120), January 2019, www.transnational-dispute-management.com.
38 www.highviewpower.com/news_announcement/world-first-liquid-air-energy-storage-plant/.
39 https://gettingthedealthrough.com/area/12/jurisdiction/35/electricity-regulation-korea/.
40 www.greentechmedia.com/articles/read/five-predictions-for-the-global-energy-storage-market-in-2019.
41 http://memr.gov.jo/EchoBusV3.0/SystemAssets/PDFs/AR/Tenders/Request%20REOI.pdf.
42 www.memr.gov.jo/EchoBusV3.0/SystemAssets/PDFs/AR/Tenders/waleedlist1.pdf.
43 https://leginfo.legislature.ca.gov/faces/billNavClient.xhtml?bill_id=200920100AB2514.
44 http://docs.cpuc.ca.gov/PublishedDocs/Published/G000/M079/K171/79171502.PDF.
45 https://leginfo.legislature.ca.gov/faces/billNavClient.xhtml?bill_id=201720180SB100.
46 https://leginfo.legislature.ca.gov/faces/billTextClient.xhtml?bill_id=201720180SB700.
47 https://leginfo.legislature.ca.gov/faces/billTextClient.xhtml?bill_id=201720180SB1369.
48 www.ferc.gov/whats-new/comm-meet/2018/021518/E-1.pdf?csrt=4293048832694754061.
49 www.nationalgrideso.com/document/84261/download.
50 www.nationalgrideso.com/insights/future-balancing-services.
51 http://docs.cpuc.ca.gov/PublishedDocs/Published/G000/M079/K533/79533378.PDF.
52 http://documents.dps.ny.gov/public/Common/ViewDoc.aspx?DocRefId=%7b0B599D87-445B-4197-9815-24C27623A6A0%7d.
53 Legislative Decree 93/2011.
54 www.ofgem.gov.uk/publications-and-updates/decision-enabling-competitive-deployment-storage-flexible-energy-system-changes-electricity-distribution-licence.
55 https://eur-lex.europa.eu/resource.html?uri=cellar:c7e47f46-faa4-11e6-8a35-01aa75ed71a1.0014.02/DOC_1&format=PDF.
56 www.tennet.eu/electricity-market/ancillary-services/.
57 www.nationalgrideso.com/balancing-services/frequency-response-services/enhanced-frequency-response-efr?overview.
58 www.nationalgrideso.com/sites/eso/files/documents/8589940795-System%20Needs%20and%20Product%20Strategy%20-%20Final.pdf.
59 https://electricity.network-codes.eu/network_codes/eb/terre/.
60 www.entsoe.eu/network_codes/eb/.
61 https://electricity.network-codes.eu/network_codes/eb/terre/.
62 www.entsoe.eu/network_codes/eb/mari/.
63 www.energy-storage.news/news/expanded-15mwh-german-battery-park-demonstrates-successful-black-start.
64 www.energy-storage.news/news/california-batterys-black-start-capability-hailed-as-major-accomplishment-i.
65 Section 4(4) of the Electricity Act 1989.
66 https://theenergyst.com/energy-storage-boom-uk-power-networks-receives-12gw-of-connection-applications/.
68 www.edf.fr/en/the-edf-group/dedicated-sections/journalists/all-press-releases/edf-announces-the-electricity-storage-plan-to-become-the-leader-in-europe-by-2035.
69 www.statkraft.co.uk/media/news/2018/statkraft-acquires-wind-development-business-in-ireland-and-the-uk/.
70 www.enel.com/media/press/d/2017/05/enel-buys-a-stand-alone-battery-energy-storage-project-in-uk.
71 www.vitol.com/press-release-vitol-company-vpi-immingham-invests-uk-energy-storage/.
72 https://uk.reuters.com/article/us-aes-siemens-batteries/aes-and-siemens-create-joint-venture-for-energy-storage-idUKKBN19W0Z8.
73 www.ft.com/content/086fe32e-6014-11e7-8814-0ac7eb84e5f1.
74 www.energy-storage.news/news/shell-invests-in-sonnen-to-drive-distributed-energy-aims-forward.
75 www.shell.com/media/news-and-media-releases/2019/smart-energy-storage-systems.html.
76 https://susi-partners.ch/news/146/64/SUSI-Energy-Storage-Fund-reaches-final-closing-at-EUR-252m/.
77 www.lseg.com/markets-products-and-services/our-markets/london-stock-exchange/equities-markets/raising-equity-finance/market-open-ceremony/london-stock-exchange-welcomes-gore-street-energy-storage-fund-plc.
78 www.energy-storage.news/news/uk-energy-storage-fund-prepares-to-hit-london-stock-exchange-with-us130m-ra.
79 www.energy-storage.news/news/california-wildfires-bankruptcy-create-uncertainty-over-pges-renewables-and.
80 https://curia.europa.eu/jcms/upload/docs/application/pdf/2018-11/cp180178en.pdf.
81 www.energy-storage.news/news/cefcs-unsubsidised-debt-finance-of-south-australia-wind-plus-storage-is-imp.
82 www.moodys.com/research/Moodys-As-cost-declines-and-regulatory-support-boost-demand-for—PR_381076.
83 www.macquarie.com/uk/about/newsroom/2019/green-investment-group-announces-debt-financing-of-california-battery-storage-project/.
84 www.santandercb.co.uk/insight-and-events/news/battery-energy-storage-solutions-ltd-secures-significant-funding-reach-100mw.

About the author

Louise Dalton
Senior associate, CMS
louise.dalton@cms-cmno.com

Louise Dalton is a senior associate in the energy practice at CMS in London, specialising in the electricity sector. She advises developers, investors, lenders, networks, suppliers and energy users on regulatory and commercial matters in the United Kingdom and internationally.

She has a focus on legal issues for flexible and emerging technologies, including energy storage, electric vehicles, peaking power and demand side response. In particular, she advises on a wide variety of storage developments comprising various technologies, including standalone, co-located and behind-the-meter projects. She regularly addresses first of a kind issues, such as the regulatory treatment of storage pursuant to legislation, regulations and industry codes, storage revenue stream matters and associated documentation.

Ms Dalton leads CMS's energy storage initiative, including coordinating the publication of the multi-jurisdictional *CMS Energy Storage E-Guide* in 2016 and *Storage and renewables: The next frontier on co-location issues* in 2018. She speaks regularly at industry events on storage and related matters and is a member of the Renewable Energy Association's Energy Storage Steering Group.

About Globe Law and Business

Globe Law and Business was established in 2005, and from the very beginning, we set out to create law books which are sufficiently high level to be of real use to the experienced professional, yet still accessible and easy to navigate. Most of our authors are drawn from Magic Circle and other top commercial firms, both in the UK and internationally.

Our titles are carefully produced, with the utmost attention paid to editorial, design and production processes. We hope this results in high-quality books which are easy to read, and a pleasure to own. All our new books are also available as ebooks, which are compatible with most desktop, laptop and tablet devices.

We have recently expanded our portfolio to include a new range of journals, Special Reports and Good Practice Guides, available both digitally and in hard copy format, and produced to the same high standards as our books.

We'd very much like to hear from you with your thoughts and ideas for improving what we offer. Please do feel free to email me on sian@globelawandbusiness.com with your views.

Sian O'Neill, managing director